Mummy Dearest

Mummy Dearest

How Two Guys in a Potato Chip Truck
Changed the Way the Living See the Dead

RON BECKETT & JERRY CONLOGUE

WITH MARK STEWART & MIKE KENNEDY

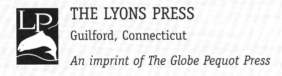

THE LYONS PRESS
Guilford, Connecticut
An imprint of The Globe Pequot Press

The Lyons Press is an imprint of The Globe Pequot Press.

10   9   8   7   6   5   4   3   2   1

Printed in the United States of America

Book design by Sheryl Kober

Library of Congress Cataloging-in-Publication Data

Becket, Ron, 1953–
   Mummy dearest : how two guys in a potato chip truck changed the way the living see the dead / Ron Becket and Jerry Conlogue.
      p. cm.
   ISBN 1-59228-544-9 (trade paper)
   1. Mummies. I. Conlogue, Jerry, 1947– II. Title.
GN293.B43 2005
393'.3—dc22

                                                                    2005010672

Dedications

JERRY

To Shar Walbaum, who believed in me and encouraged me to make my first trip to Peru. To my son, Byron, and daughter, Keanu (aka Ali), for enduring their father's eccentricities. And to the countless radiographers, such as Ray Gagnon and Marty Ricart, who were my teachers and enabled me to meet the imaging challenges I encountered on Mummy Road Show.

RON

First and foremost, to my children, Matthew, Paul, and Julie. Life is so full of wonder and beauty, none of which compares to you. I am glad that each of you got to experience a small part of Mummy Road Show *as soundman, cameraperson, note taker, photo documenter, and re-enactment extras. Never was there a time in all of my travels that you three kids weren't right there in my mind and in my heart. To my partner, friend, and wife, Kathy, whose love and constant encouragement has given me the courage to fly. To my grandfather, Ralph "Buster" Beckett, who inspired me to consider science and research and to love and respect life. And to my parents, Howard and Terry Beckett, who have never failed in their enthusiastic love and support of me as I chased many dreams throughout my life.*

Contents

Introduction

What happens to you when you die? Does your spirit migrate to some new realm? Are you reunited with loved ones? Do you meet God? Or is that just it? The answer is that nobody knows, because nobody has ever come back to tell us. That, we believe, is why people of all ages, from all parts of the world and from every walk of life, are so intrigued by, and curious about, mummies. Basically, they are the only dead folks still hanging around.

We—Ron Beckett and Jerry Conlogue, that is—are fascinated by mummies, too. As scientists in the health-care field, however, we tend to focus on a slightly different mystery. The mystery of life. There is so much we do not understand about the business of living and dying—not just in ancient times, but all the way up to the fairly recent past. Yet, if you know what to look for, mummies can fill in a lot of the blanks.

Mummy investigation is our hobby, but for three years you could say it was our job. At least our part-time job. We were the hosts of a television series called *Mummy Road Show*, which aired worldwide on the National Geographic Channel US and National Geographic Channels International. The show did a wonderful job of conveying our curiosity and passion about mummies, and showcased techniques we had developed together which, hopefully, advanced the science and methodology of this growing field.

Besides making us cable-TV rock stars (just kidding), *Mummy Road Show* really expanded our horizons. The program enabled us to study a wider range of mummies than either of us had ever imagined, and helped us make incredible professional and personal connections, which continue to enhance both our lives and careers to this day. The show itself struck a fine balance between education and entertainment (a tribute to its producer, Engel Brothers Media), and cultivated a devoted following.

We still hear from fans of *Mummy Road Show*. Unlike the human body, once a television program's natural life ends, we *do* know where it

Life in the fast lane! Here we are—Ron Beckett (left) and Jerry Conlogue (right)—stranded with a broken-down potato chip truck full of equipment, waiting for a mechanic to show up.

goes: syndication. We are very big in Australia and India, where the show appeared as a first-run series.

Another aspect of television that we came to understand is that, because you have about twenty-three minutes to tell a story (and often a compressed amount of time in which to film an episode), some terrific stuff ends up on the editing-room floor. Also, a lot of the best stories simply never found their way onto film. What makes for a great yarn, we found, does not always make for good TV.

Mummy Dearest **is really a love letter to the camaraderie and spirit of discovery shared by the people in this field.**

That is not the reason we wrote this book, however. Although devotees of *Mummy Road Show* may love all the juicy behind-the-scenes stories, *Mummy Dearest* is really a love letter to the camaraderie and spirit of discovery shared by the people in this field. As you will see, we define "field" very broadly. People who are passionate about mummies, who devote their lives to studying and caring for them, run the gamut. Our network of contacts and colleagues includes world-renowned anthropologists, museum directors, under-the-radar collectors, carnival operators, and more than a few people who have known little more luxury in their lives than a straw mat on a dirt floor. (We fit in there somewhere, trust us.)

To know these individuals is to gain a deeper and more profound understanding of the culture of mummies. As you move through the pages of this book, you will get to know them, too. They are as interesting as the mummies themselves sometimes, and we are proud to have met them and to count ourselves among them. The bond we share—whether we work out of a paneled office at the British Museum, a cave in the Philippines, or the cluttered fifteen-by-fifteen university offices we call home—is that each of us can provide a crucial piece to the big picture, which grows more complete, and more complex, with each passing day.

So hop on into our old potato chip truck, find yourself a spot among the X-ray units, endoscopy gear, and other equipment, set up your folding chair, and hang on. There are no limits to this ride. And no seat belts, either.

Acknowledgments

In the pages of this book, you will get to know many of the people who made *Mummy Road Show* such a success. Needless to say, there are many more you won't. We would like to take this opportunity to recognize the contributions made to the show, this book, and to our colleagues in this fascinating field of study.

Let's start by thanking all of the dozens of wonderful people at Engel Brothers Media (now Engel Entertainment) that made *Mummy Road Show* a terrific adventure and great television. Our special thanks go to executive producer and producer Larry Engel (who also directed and was the cinematographer on most of the episodes) for his tireless efforts in getting the best out of us; Mary Olive Smith, series producer and producer, for overseeing story development and for piecing together forty episodes over three seasons; Amy Bucher, who not only produced some of the best episodes of *Mummy Road Show*, but first introduced us to Engel Brothers Media while directing and producing another Engel Brothers Media film in Peru called *Desert Mummies of Peru*; Alana Campbell, for sticking with us and working her butt off and going from production assistant to co-producer over the three years of *Mummy Road Show* in the process; and Susan Lee for tirelessly and often thanklessly undertaking the role of production manager for the entire run of the *Mummy Road Show* series and keeping her good humor throughout.

Then there are our colleagues in the world of science and medicine who must be thanked, including Sonia Gúillen, bioanthropologist and director of Centro Mallqui of The Bioanthropology Foundation Peru, who introduced us into the field of mummy investigation and who remained with us through *Mummy Road Show*; Andrew Nelson, physical anthropologist, University of Western Ontario, for the connections and inspiration he provided to us and to the Engel Brothers Media crew throughout the series; Larry Cartmell, pathologist, Valley View Regional Hospital, Ada, Oklahoma, for his endless behind-the-scenes contributions analyzing mummy tissues in the lab; Arthur Aufderheide,

MD, the granddaddy of mummy paleopathology and our unofficial research advisor, who kindly offered answers to our queries and inquiries; Ron Martin, associate professor, Department of Chemistry, University of Western Ontario, for his guidance and supervision in mummy tissue analysis; and the staff at Trace Elements Laboratory, University Hospital, London Health Sciences Centre, in London, Ontario, Canada.

We would also like to thank all the folks at National Geographic Channels International and the National Geographic Channel U.S., which together aired *Mummy Road Show* in over 135 nations around the world. We want to make special note of Martha Conboy and John Bowman, series executive producers for National Geographic Channels International and the National Geographic Channel U.S., both of whom gave the Engel Brothers Media team enough freedom and guidance to produce a quality series for three seasons; and Tracy Beckett and Janet Han Vissering both of whom had the foresight in their executive capacities at National Geographic Channels International to champion the series as soon as they received the treatment and pitch tape from Engel Brothers Media and were instrumental in getting the series green-lighted.

Finally, we would like to thank Heidi Reavis, attorney extraordinaire, who supported both the *Mummy Road Show* television series and this book with her legal, emotional and creative support and made countless problems disappear—quietly and expertly; Mark Stewart and Mike Kennedy for making this book as good as it is through their creativity and whip-cracking; and Steven Engel, the head of Engel Entertainment and Engel Brothers Media, executive producer (and sometimes producer) of *Mummy Road Show*, who had the intelligence to marry Heidi Reavis and the keen eye to recognize our potential as co-hosts and co-authors, thus making both the television series and this book a joyous reality for the two of us.

Chapter One
Road Warriors

RON BECKETT

Born: January 3, 1953

Height: 5-8

Weight: 168 (Okay, so it may sometimes sneak above 170)

Hair: Lots

Hobbies: Playing music, carpentry, hiking, and kayaking

Turn-ons: Sunsets, warm desert nights, my kids, a good Martin guitar and "m'lady of the desert"

Turn-offs: Politicians in general, and anyone stuffy: faculty, physicians, lawyers, etc.

I'm at my best: Always

I'm at my worst: Never (except maybe administrative tasks)

Secret ambition: I'm living it

Preferred method of burial: Funeral pyre in the open desert

If I'm uncovered as a mummy one hundred years from now, they'll make up this story about me: Researchers will have a heck of a time figuring out what was important to me, because I will be accompanied by such a variety of burial goods. Also, they will see from my bones that I have lived life not as a spectator but as an active participant, from my artificial hip and many healed bone fractures! If they're able to somehow read my brain, they will discover that I was . . . well . . . "out there!"

I was born in the shade of a saguaro, raised by wild coyotes. (Just kidding.) I was born in Yuma, Arizona, a town in the state's southwest corner bordering California, and twenty-three miles from Mexico. There was only one paved street there when I was young, so it was really an amazing place for a kid to grow up. (Actually, I found out later that there were other paved streets in another part of town.) Our neighborhood was on the edge of town, but Yuma grew rapidly, and paved roads did eventually come in. Even so, when we would go to visit my cousins in Chicago, they would ask us if we had running water or a TV. It must have seemed very primitive to them. To us, however, the whole world was our playground.

The cultural diversity of the people who lived in Yuma was just amazing. You had families like the Becketts, you had Mexican-American families, Asian-American families, and African-American families, and you had indigenous people, some of whom probably could trace their ancestry in this region to a time before Europeans set foot on North American soil. All of my summer jobs involved working in the agricultural fields with these great people. What a fantastic way to be exposed to other cultures! I remember trading my peanut butter and jelly sandwiches for potato burritos. Today, I look for cultural similarities because of these experiences.

My parents, Terry and Howard, my brother Greg, and my sister Donna all still live in Arizona. My mother's family is from Chicago, a generation removed from Croatia. My father's family came to the Southwest when his father became a researcher working for the agricultural departments of both the United States and Mexico. He set up something that came to be called the Monkey Farm near Yuma. My father was interested in agriculture, too. Research and science were always a part of our family, and although I didn't realize it back then, obviously this had a great impact on me.

My father was studying agriculture at the University of Arizona when he decided to join his brother in the refrigeration business. Later, he became an insurance agent. He believed so passionately in insurance as a retirement plan that he would go down to the poorer sections of Yuma and get people started on a policy. If a customer could not make a payment, my father would pay their premiums from the overrides he made as district manager. The company caught wind of this and ordered him to stop. He disagreed, and quit selling insurance.

Because of my dad's job, we had a pretty nice home at the time, with an in-ground swimming pool. My father explained why he quit, and that it was important to stick to your principles, even if it meant moving to a smaller house. I was so impressed with his decision to do that. Later, he got into the roofing and insulation business, and did fine. He was an excellent carpenter. Back when he was in the insurance business, he actually built his own office. He also built our first home. I loved working with him on those kinds of projects.

I learned a lot from my dad. I learned how to improvise. I learned there were no limitations. He would let me get up there on the roof with him as a kid, and I marveled at how he could make any piece of wood work in the right place. If he ran out of nails, he straightened the bent ones. To this day, I still embrace this philosophy: Think and make it work.

If he ran out of nails, he straightened the bent ones. To this day, I still embrace this philosophy: Think and make it work.

I credit both of my parents with giving me the ability to make decisions. I guess I would characterize their approach as "gentle guidance." If I asked, "Mom, can I do this?", she would respond, "What do *you* think?" I remember her telling me once, "We've shown you right and wrong; now it's *your* turn to make those decisions." I was only about eleven or twelve when she said that. I thought that was really neat.

With the freedom to make choices, of course, came responsibility for those choices, as well as an expectation of appropriate behavior. This fostered an environment of mutual respect and helped me develop the same compassion and love my parents showed to us, and to other people, as well.

My father, for example, used to think Paul Simon's song, "The Sound of Silence," should be our national anthem. He really worried about the homeless. We had an El Camino, and once I remember putting the groceries in the back and stopping at another store. I said, "Dad, do you want to lock these

up?" He said, "No. If someone takes the food, they need it worse than we do."
He was that kind of guy.

Growing up in Yuma was a blast. Greg and I had nothing but fun. Everyday it was the same thing: What are we gonna do? We're gonna have fun!

We'd play in the desert, we'd jump off the neighbor's roof and get
hollered at, take our bikes out into the desert and ride them straight down
hills until they fell apart. We were kind of wild, kind of adventuresome. I'd
say my brother was pretty much my best friend. We hung out together a lot,
so we didn't torment my older sister too much. She was an absolute sweet-
heart, a Homecoming Queen! We are all still really close today.

My circle of friends began to grow when I got into high school. I was involved in sports and the music department, so I had an eclectic group of
friends. That has been the case ever since, and what is kind of cool is that
my own kids are that way now, too. They are not judgmental about people.
They take people at face value and try to pull the good out of them.

I played football in junior high and for four years at Yuma High School.
I was a lightweight, between 135 and 140 pounds, but I was a first-string
linebacker, which tells you a little about me and a lot about the size of our
school system. We would play schools from Phoenix, which was about three
hours away, and I had to tackle these 230-pound fullbacks who busted
through our line and came full bore at me. I must have felt like a fly on their
legs. Yeah, it hurt, but I never backed down.

Our coaches were really good because everybody got to play. They really
believed in sportsmanship. Obviously, we did not win too many games, despite
our team name. We weren't the Bulldogs or Bobcats or Eagles. We were—get
this—the Yuma Criminals. Seriously, our mascot was this really nasty-looking
convict.

Still, the football team was a unique experience that made me appreciate how important individual teachers can be in giving kids a good education. Among my favorite teachers at Yuma High was a music teacher, Taylor
McBride. I was already into music before high school, but he had a way of
instilling confidence in his students. He would say, "I expect your best—I

don't expect you to be great, but I expect your best." When you gave him your best and he recognized it, you had a good, warm feeling. What Taylor McBride was really teaching us was to not be afraid of failure and to work up to your abilities. Don't worry that someone sings better than you, or plays an instrument better; be good at what you can do, and have fun with it. We had a blast with that guy.

What a contrast this was to my earlier educational experiences. I attended Alice Byrne Elementary School through third grade, and apparently, I just didn't get it. I was always coloring something incorrectly or committing some other infraction because my most vivid memory is of a 900-year-old teacher constantly whacking me on the knuckles with a ruler. From there I went to St. Francis, a parochial school in Yuma. There were some good teachers there, and some not so good. My music teacher was another knuckle-cracker. I learned piano from her, and *Whack!* if you didn't play it right. That was not fun, and music was supposed to be fun, so I stopped taking piano.

> **My music teacher was another knuckle-cracker. I learned piano from her, and *Whack!* if you didn't play it right.**

Did I bring some of this on myself? I don't know if I was the class clown, but I did have fun, and I did get a little bit of a talking-to from many of my teachers. My mother said that one time in the hospital after I was born, she was worried because the nurses weren't bringing me to her for a feeding. She discovered that they were having too much fun with me. Apparently I was laughing and cutting it up even as a newborn.

I could also be stubborn. My first protest was on the basketball court at St. Francis. The guys held a sit-in on the court because the girls were playing volleyball on it. They did have another volleyball court, but we couldn't play basketball anywhere except the basketball court. So we thought that was unfair. Our protest made the head nun cry.

All told, I did pretty well in school and got pretty good grades. At Yuma High, I started goofing off a bit. There were times when I felt I was not being challenged, and then there were times when I just had too many ideas whizzing around my brain. But I continued my education, enrolling at Northern Arizona University in 1970 as a music major, with the goal of becoming a music teacher.

Music was my passion at this point. I had started with piano as a child and moved to guitar in 1964, right after the Beatles appeared on *The Ed Sullivan Show*. I remember my mother rushing to get me to watch with her and my dad. My parents loved the Beatles as much I as did. They were mesmerizing. I already wanted to play guitar because my sister's boyfriend played. My uncle Johnny played a little bit, too. The Beatles just sealed the deal. I got a cheap old guitar and took about three months of lessons from Leonard Jones, who played with a real Chet Atkins style. From then on, I had the rudiments. I began to self-teach, and took up the bass guitar, started learning music theory, and played in rock bands at high school and later in college. I was in a great soul band, too. In fact, at one point in college I was in eight different music groups at the same time!

I did well in my music classes at Northern Arizona, but not so well in my other classes. I took a biology course, which included a biology lab, but I completely forgot about the lab class and literally missed the whole semester. I guess you could say I still had a lot to learn about being a student. I decided I should take some time off, do some traveling, learn about the world outside Arizona, and play my music.

Up with People

Remember Up with People? That upbeat, high-energy musical show? Well, for two years I was a member of that perpetually cheerful organization. It was a wonderful experience on a number of levels, and a real eye-opener, too. Here I was, this boy from Yuma, flying off to LaGuardia Airport in New York City and meeting up with a group of strangers traveling the world and performing in places like Australia, Spain, Germany, Mexico, and Central Europe. That was the kind of confidence and sense of adventure my parents had instilled in me.

We spent several months in Hawaii, and I had an important experience there. One of my bandmates was from Belgium, and he took me to the leper colony on Kalaupapa, where a famous Belgian priest, Father Damien, served as the colony's spiritual guide. So we hung out with the lepers for a while. We played music, toured Kalaupapa, had a pig roast. How many people can say that?

These lepers should have been dead, but in their lifetime new medicines were developed that arrested the disease. This gave them a beautiful outlook on life, living in the moment and appreciating every second they had. It was a great lesson for me.

The kids in Up with People were friendly, talented, and enthusiastic. I think that came through in the performances—so much so that people kind of poked fun at us at times. We didn't care. We were having a great time. Where I had a problem was that the organization was constantly hammering into us how special we were. I felt special because I got to do this great thing. But I didn't believe I was special in a superior way. I think some of the kids believed this was the case, however, and Up with People did nothing to discourage this view. That rubbed me the wrong way.

In the end, I also began to question the need for the organization itself. Maybe it was the hippie in me, but I thought all these nice young people from all over the world were getting a skewed vision of themselves, and wondered how that would affect them later in life, when there was no one around to get them pumped up. I also suspected someone was floating around on a yacht somewhere thanks to the money we generated.

When I returned to Arizona a couple of years later, I began to think, *What can I do? How can I contribute to society?* Music was still an option, and I thought about teaching. But my buddies and I had done some rappelling and technical climbing, so we offered our skills to the search-and-rescue team in Tucson. I was actually deputized as a search-and-rescue volunteer. There wasn't much rescuing in those mountains, though. It was mostly body recovery. Still, the experience introduced me to health care. I thought, *health care, this is a great way to help people.*

I looked at the different professions, and I chose a health science—respiratory care—because I could be involved with patients at a critical time, when their breathing and circulation were in jeopardy. I learned the ABCs of CPR, and I got to work on transport teams. It was very exciting. There was a lot of technology, and also a lot of people skills involved. It was perfect for me.

I enrolled at Pima Community College in Tucson to earn an associate's degree, and where I had been an academic disaster a couple of years earlier, now I was unstoppable, a straight-A student. I began working in the respiratory care field, and then fulfilled my ambition to become a teacher. I actually taught at the college I had attended, becoming a clinical instructor at the hospital. Within three months, I was promoted to supervisor. I had a lot of responsibility early in my career, and I loved it. Since then I have always seemed to be working at least two jobs at a time.

Being in hospital middle management and teaching at the same time enabled me to look at the health-care field from two different angles. Eventually,

I decided to move my career toward teaching full-time. I got a BA in health service administration from the University of Phoenix, which offered an external degree program that was perfect for someone with two jobs. Then I moved to the east coast and began teaching. I completed my master's in education at Rhode Island College while teaching at the Community College of Rhode Island, and later earned a doctorate in adult education from the University of Connecticut.

The Rhode Island job was good, but I never ended up with the salary that they promised me. When a position teaching respiratory care sciences at Quinnipiac University opened up in 1985, I grabbed it. I began teaching anatomy and physiology labs, and teaching pathophysiology as well.

Jerry and I met in 1992. We needed a new program director for Diagnostic Imaging. I was department chairman, and Jerry's CV came across my desk. I really liked the diversity of his background. He didn't just do diagnostic imaging of humans—he did everything from porpoise flippers to whale heads (though as I recall, he may have left that one off his résumé). Anyway, this was what I wanted my students exposed to. I picked up Jerry at the airport, and it was kind of like we were friends already. We enjoyed a similar outlook on life. Slightly different styles, maybe, but we hit it off. I thought, *Okay, this is the guy*. So we hired him. In working together we found that we share the conviction that life should not be a spectator sport. People have the ability within them to do anything. It's a matter of courage. It's a matter of confidence. It's a matter of understanding how you can get to where you want to be. This is something we both try to convey to our students at Quinnipiac.

One great experience Jerry and I shared was being involved with the necropsy (an autopsy on an animal) of a 42-foot finback whale that had died among the pilings in New Haven Harbor. The whale was hauled up to the town dump where several research teams tried to determine why this great sea mammal had perished. Jerry had worked with whales before, so I learned a great deal that day. One of our colleagues did research on whale phonation, so we were able to remove the whale larynx for the phonation research. We "de-fleshed" the larynx in the snow back at Quinnipiac, and it looked a little like a battle scene, with whale blood all over the place. Unfortunately, we

Jerry sizes up a beached whale.

didn't learn why the whale had died, but I learned something that Jerry already knew: Nothing in this world smells quite like whale blubber. It permeates everything. The odor penetrates into your skin, and you carry the stench for days. We had to throw away our clothes—they stunk!

Jerry also taught me how to model animal structures, and I began researching the anatomy of the pig lung, an organ that may one day be transplanted into humans. I got fresh pig lungs from the local slaughterhouse—now there's a place you don't want to go on a first date!—and brought them to my lab. There my students and I injected silicone caulking into the airways. After gently boiling away the tissue (which also reeked), we had a great model of the airway structure remaining in the lung. This research continues to this day.

Jerry throws himself into his work.

In 1996, Gretchen Warden of the Mütter Museum in Philadelphia got Jerry involved with Andrew Nelson, a Canadian scientist who would soon become our colleague. Andrew was planning a trip to Peru to study mummies, and Jerry agreed to accompany him. Prior to their departure, we were in my office, and Jerry mentioned that sometimes there would be shadows on mummy X-rays that were impossible to decipher. I was all ears. Ever since Jerry had relayed me the story of the mummy at the Mütter known as the Soap Lady, mummies had intrigued me. I told Jerry that if there was an opening, I was pretty good with an endoscope; perhaps I could solve the inner mysteries that the X-rays couldn't.

That was our eureka moment.

We made a fake mummy bundle, placing a pen and paper clip at various levels. He x-rayed them from different angles. I scoped around, and sure enough, the combination worked perfectly. I began joining Jerry on excursions to museums and searching for literature on applications of endoscopy that might be relevant. Everything I found followed the medical model, where you have a target and go in for a biopsy. Mine was more of a survey and investigation model, which was uncharted territory. So I developed techniques as I went, and found that they really complemented all the things Jerry could do with X-rays.

When *Mummy Road Show* came along—give Engel Brothers Media credit for selling the idea for the series to National Geographic Channels—it was tailor-made for my interests and outlook. The prospect of travel was exciting to me, but more than that, there was the allure of piecing together the past, understanding and respecting people from different cultures, and finding solutions where others had only seen limitations. The concept for the show seemed to embody everything my parents had taught me.

The endoscope in action during "Mummy in Vegas." That's me on the right.

The pay was going to be lousy, the time demands would be unreasonable, and the working conditions atrocious at times. There were just two things we wanted to know: *Where do we sign?*, and *When do we start?*

And so the adventure began.

JERRY CONLOGUE

Born: December 19, 1947

Height: 5-10

Weight: 225

Hair: Long, and in a ponytail

Hobbies: Poking around used bookstores, searching for abandoned quarries, and taking X-rays of unconventional things like flowers and insects

Turn-ons: A challenge (It's hard for me to say no)

Turn-offs: Someone who tells me a task is impossible; someone who disregards the feelings of others

I'm at my best: When I don't have a time limit

I'm at my worst: Memorizing lines or early in the morning

Secret ambition: Learn to fly an Ultralite

Preferred method of burial: With my '50s-era Picker Mobile Army Field Unit

If I'm uncovered as a mummy one hundred years from now, they'll make up this story about me: If it's an entrepreneur trying to make a profit off my carcass, it's going to be outrageous, and it can't be like the stories about Sylvester, Hazel Farris, or Marie O'Day. The outlaws today are in some ways more diabolical. It will go something like this: I was a notorious CEO of some robber-baron corporation, responsible for the loss of thousands of jobs and the life savings of all my employees. I put dictators and puppet governments in place so that I could pillage the wealth of a third world country.

I like the unknown, and I love a challenge. Every now and then I say to Ron, "We have no idea what we'll be doing a year from now"—which always brings a smile to my face. When the opportunity came along to do *Mummy Road Show*, I grabbed it. I had no idea what to expect, but that was a bonus.

My childhood couldn't have been more different than Ron's. In fact, I tend not to discuss my past, because the more I look back on it, the more I want to put it behind me. But writing this book was an interesting experience. It brought back so many memories—good and bad.

I was born in Meriden, Connecticut, which is about midway between Hartford and New Haven. My mother lost five children, either by miscarriage or shortly after birth. I was an only child, so my parents were very protective of me. At the age of five, I was diagnosed with rheumatic fever (the first of several childhood illnesses), so my earliest recollections involve a lot of doctors, hospitals, and home tutoring instead of playgrounds, school chums, and the all-American family vacations you might picture when thinking of the 1950s and '60s. My doctor was very conservative, and he maintained that complete bed rest was the most prudent way to manage my illness. Unfortunately, I think the isolation made me more susceptible to conditions like pneumonia, which resulted in more hospitalizations.

My parents, Jennie and Paul, both worked in Meriden, which for decades had been a manufacturing center for silver and stainless-steel flatware and dinnerware. They were employed in this industry for many years, but business began to tail off during my childhood years, and the kind of factory jobs they had held started to disappear. Meriden was also home to a General Motors plant, and when it closed down the town was hit especially hard. People think of the 1950s and '60s as boom years, but in Connecticut this definitely was not the case. It was really difficult on my family, so my father always worked two jobs. Since my mother worked, too, I was raised to a significant degree by my grandmother, Rose Vittorio.

I grew up believing that you always try your best, work your hardest, and never just sit back and wait for things to come to you.

Needless to say, my parents instilled a great work ethic in me. I grew up believing that you always try your best, work your hardest, and never just sit back and wait for things to come to you. Long before *Just Do It* became Nike's marketing mantra, it was a way of life in the Conlogue home. At the same time, my parents also had great compassion for others. If someone needed help or support, you provided what you could, no questions asked. These two forces continue to play a major role in my work and my life.

I did not form a lot of friendships with people my own age because I was around them so infrequently. Occasionally, I would be sent to school for a week or so, and then pulled out. This was intended to give me a taste of a regular childhood, but I'm not certain it was such a hot idea. The kids weren't sure how to treat me, and they could also be cruel, which sometimes made me feel even more isolated.

In a funny way, I was probably more comfortable in a hospital than a school as a child, despite the fact that hospitals could be scary places. I remember going in for a tonsillectomy at the age of five and waking up the night before my operation to a big commotion in the next bed. This was when they kept kids in these glass cubicles, so I peered through the window and watched as doctors sliced open this boy's throat—he was unconscious—and put a tube in. Looking back, I realize now they were performing a tracheotomy, but to my five-year-old eyes, everything told me this was a tonsillectomy. I understood that they were taking something out of my throat and that I would be asleep while it happened. Naturally, I assumed they were coming for *me* next.

After working on the boy for a while, suddenly everyone stopped. Then they pulled the sheet over his head. Well, I knew what that meant. It was a truly frightening experience. When they came the next day to get me to remove my tonsils, it took about five of them to hold me down before they gave me the ether to knock me out.

The kids I met during my subsequent hospital stays were incredible. Many were dealing with devastating illnesses with such courage. It made me feel lucky that I was only in there for a week or two. I developed really close relationships with these kids in the short time we knew each other, because we were sharing something that no one else could understand. We all sensed that we were missing something precious. We weren't going to ride bicycles or climb trees or play football. We weren't going to get to be kids.

Years later, when I was twenty-five, I went to Sears and bought a three-speed bike. I found an empty parking lot and taught myself how to ride. I would wobble a few feet and then fall over, just like Arte Johnson on the old

Laugh-In television show. I look back and laugh at myself now—which is also how people watching me that day reacted—but at the time it was kind of a profound experience.

The feeling that we sick kids were not like other kids was underscored every time we interacted with a doctor or nurse. Back in the 1950s, remember, there was a whole different attitude toward children in the hospital. They didn't tell you anything. People would say, *This isn't going to hurt*—but we knew better. Like hell it wasn't going to hurt. It hurt—what the hell are you talking about? They would say, *Look the other way*. Then they would do something terrible to you. Not that I mistrusted adults. On the contrary; I felt I connected more with them because these were the people with whom I interacted most often. My parents, my grandmother, doctors, hospital staff, the wonderful tutors the state sent to our home—these were the people who populated my world.

I was certified disease-free and unleashed on the general population at the age of twelve. My first school year was really difficult, because I felt I had so little in common with kids my age. I had never played sports or engaged in any kind of strenuous physical activity. Prepubescent boys can be incredibly cruel to someone who is different, and I experienced this right through junior high and into high school.

Compounding my misery was the fact that there simply wasn't anyone you could talk to back then. You did not have school counselors or other resources, which meant you had to work through stuff on your own. In the late 1960s, a lot of those kids became young adults and turned to drugs as an escape. I lost a lot of friends to overdoses over the years, and that's something that stays with you a long, long time.

One of my escapes was books. Unfortunately, I probably had—with the criteria used today—a learning disorder. I loved to read, but I could not get through an entire book until my junior year in high school. It was something I simply adjusted to, and I learned to live with it. Again, there was no one available to diagnose or address the problem, so it was just another thing I had to work out on my own.

I suppose you could say that my professional career began at the age of sixteen, when I got a job in the meat section of the local supermarket. At that point, my aspiration was to become a meat cutter. By the time I graduated from Maloney High, however, I had set my sights somewhat higher. I wanted to work in a hospital. Having been a patient all those years had made me comfortable in this environment, and treating patients with more sensitivity and compassion was important to me. To this day, I still identify with young people who are chronically ill or stuck in a hospital. When I get someone in my class with a disability, I also feel a kind of kinship, although I don't give them any advantage over the other students.

I got into the X-ray program at Hartford Hospital at the age of seventeen, and I worked nights as an ambulance attendant. I had no money, so this was an ideal gig. I could sleep there for free and when a call came in, I would jump in the ambulance and get paid two bucks for everyone we transported to the hospital. That was decent pay in 1965. Our boss also paid us to take pictures at accident scenes, which he sold to the highest-bidding attorney. Later, I became a lab diener, which is an individual who assists on autopsies. That is how I got started working with dead people.

> **To this day, I still identify with young people who are chronically ill or stuck in a hospital. When I get someone in my class with a disability, I also feel a kind of kinship, although I don't give them any advantage over the other students.**

Meanwhile, my eyes were being opened to the possibilities of radiology. While a student at Hartford Hospital, I met a radiographer, Bob Pooler, who showed me some of the extraordinary things X-rays could do—not to mention some of the less-than-legitimate applications. There was a parking gate that operated with an electromagnetic key. If you x-rayed the card, you could see the location of the little squares that opened the gate. He showed me how to make parking keys for people who wanted to park in the lot by using squares cut out of stainless-steel razor blades. This was one of the little cottage industries that I participated in from time to time to feed myself.

The X-ray program was a great experience for a young person. There were about twelve of us in the class of 1967, and everyone looked out for one another. I try to reproduce that camaraderie in my own classes now by encouraging students to work together in unusual ways so they can achieve this level of teamwork.

I formed a friendship with this one guy, Ron Sagram, from Trinidad. He was fifteen years older than I was, his family and friends

Ron and I examine a skull X-ray for "Egypt, California Style."

were thousands of miles away, and he had no one to hang out with. We worked together x-raying bodies in the morgue, and we found that we were both interested in pushing the envelope a bit. We borrowed some surgical specimens from the morgue and eventually we developed a technique for injecting them in a way that demonstrated circulation. There was another technologist, Charlie Maccalous, who was interested in building things. He actually took an electric mixer and put a different gear system on it, so we could slowly inject specimens. This actually led to a job at Yale University when I graduated from the program in 1967.

At Yale's radiology research lab, I met this extraordinary radiographer, Joe Sarmony. He had been a practicing doctor in Hungary in the early 1940s, and was inducted into the German army when they took over during World War II. At the end of the war he was captured by the Russians, but somehow found his way to Canada. His records were lost, so instead of repeating medical school he became a highly regarded X-ray technologist who was later recruited by Yale to run its research lab. He taught me anesthesia and surgical techniques that enabled researchers to test procedures on animals before they were used on people.

By 1968, the Vietnam War had drained enough out of the federal budget that they were making cuts in programs like the one at Yale, and I found myself out of a job. I ended up working at Bellevue Hospital in Manhattan. New York University had recently purchased Bellevue, and Senator Jacob Javits had worked out a deal where you could work there for four years instead of going to Vietnam, so I was fortunate to have gotten in.

Bellevue was an incredible place to work. There were all kinds of cases that I probably would not have seen anywhere else. One night, a riot broke out at a George Wallace rally being held at Madison Square Garden. Suddenly, we were inundated with what seemed like an unending stream of injured people. It was here that I really developed my skills as a radiographer, and learned how to improvise from some of the technologists who started out as darkroom technicians. They had learned the profession on the job, without the formal structure of the program I had recently completed. During this period I also worked part-time on weekends at the Animal Medical Center and developed an interest in veterinary medicine. Probably even more important, I realized that what I had learned about imaging humans could easily be applied to other species.

In 1970, I was invited back to Yale to take over the X-ray lab. I accepted this offer, while continuing to work at Bellevue on weekends to finish out my four-year obligation. During my second stint at Yale I went back to school, part-time, at Southern Connecticut State University and Central Connecticut State University. While I was at Central, I started working with a graduate student who was interested in looking at the structure of fishes' heads. Fish are able to smell through a system called the lateral line. It goes along the side of the body and around the head. What he wanted to do, without cutting up the fish, was to demonstrate this. I injected a contrast media (a material that shows up on X-rays), and we actually brought the fish to Bellevue and x-rayed it. Through these types of experiences, I not only started becoming interested in medical research, but also in what I call "non-traditional" applications of X-rays. It was a little more rewarding than making parking passes. By now I was taking classes at the University of Connecticut. I graduated from there in 1974, and ended up going on for my Master's degree at Quinnipiac, where I teach today.

Whale of a Tale

Y ou don't become an expert in anatomy without getting a little messy from time to time. Still, there's *messy*, and then there's a place so far *beyond* messy that you look back and wonder what the hell was going through your head at the time.

While working on a comparative neuroanatomy project at Yale, we went out and collected a number of different animals to study the vascular structure of their brains. One day, we received a call about a whale that had stranded itself on a Rhode Island beach and subsequently died. *Did we want the whale's head?*, they wanted to know. Are you kidding? I rented a chain saw and piled into my van with Alden Mead, the director of Yale's Ophthalmology Research Lab. We were there in a flash. After totally destroying the chain saw and realizing that a whale head would probably not fit in the van, we came back with a flatbed truck and an instrument called a flensing knife. We used two tow trucks to lift the severed head and backed the flatbed underneath it.

We drove the head to the Peabody Museum in New Haven for the dissection. As we discarded pieces, the smell of freshly butchered whale wafted through this very upscale neighborhood and attracted a lot of dogs, who got into the paper trash bags filled with decomposing tissue. The dogs brought chunks of greasy, foul-smelling whale meat back to some very nice homes and consumed them on some very expensive Oriental rugs. We were asked to move the whale's head.

I talked to someone at Yale about our dilemma, but without describing the magnitude of this head, and was told we could put it at Cox Cage, the sports field. We drove the flatbed onto the athletic field, tied the head to a goalpost, then put the truck in gear and drove it out from under the head. Needless to say, within a couple of days, my job was on the line at Yale. I had to move the head again.

At this point, I was finishing up graduate school at Quinnipiac. I had a project going at the time that involved the study of a parasite that can be transmitted from rats to humans. This required me to trap rats at three locations in Hartford: the city dump, the North End, and an area called Frog Hollow. The study went really well, and gave me some expertise that not a lot of people have. My advisor was very pleased with the work I was doing, which enabled me to convince him that a severed whale head would be a once-in-a-lifetime experience for Quinnipiac's biology students. If I was allowed to bring this Yale whale to campus, I told him, they could get involved with the dissection. Once again it took two tow trucks (which I charged to Triple-A) and the flatbed to move it to the soccer field at Quinnipiac.

Within a week, every raccoon, fox, and other carnivorous creature had descended from the Sleeping Giant—the forested ridge that rises over the campus—and had begun feasting on my whale head. I suddenly sensed my graduation was in jeopardy. I had to get rid of the head again.

We summoned the tow trucks (again, on Triple-A's dime) and lifted the whale head back on the flatbed. At this point, by the way, the owner of the flatbed was now very anxious to get rid of the vehicle. It seemed the oils from the decomposing whale had seeped into the truck bed. What may have been a repulsive smell to humans was irresistible to dogs living anywhere near the truck. Every imaginable breed of dog from near and far seemed to make the pilgrimage to the site. Unable to bring a souvenir home to their unsuspecting masters, they did the next best thing, and rolled around on the truck bed. Not surprisingly the price of the flatbed had come down considerably since the whale adventure began, and I could have purchased it for five hundred dollars. Our next destination was northeastern Connecticut, where a Native American gentleman had read about the head in the news and thought it would be neat to have it in his field. Unfortunately, by the time we drove up there, we learned the bank had foreclosed on his property, and we had to take it somewhere else. Thankfully, we found a home for it at Mystic Aquarium.

I kept my job at Yale, received my degree at Quinnipiac, and got into a graduate program at Washington State University, which was good, because everyone I knew in Connecticut wanted me three thousand miles away.

After Quinnipiac, I hitchhiked to Pullman, Washington, and met with the professor who wrote one of the medical entomology textbooks we used in the master's program. I convinced him to accept me into the graduate program at Washington State University. I returned to Connecticut, threw my few belongings into my van, and headed west. I had no money at this time and knew nobody out there. I earned a few bucks as a teaching assistant and found work as an X-ray technologist, but I couldn't afford an apartment and lived in that van in the parking lot for almost a year. Among the many things I learned was that it can get very cold in Washington.

Despite the hardships, this ended up being a positive experience. I learned that if I focused hard enough on something, I was going to be successful. This helped me develop a lot of confidence, and showed me that I thrived on challenges. In fact, this was one of the most enjoyable aspects of doing *Mummy Road Show*, because each episode presented us with different challenges.

Sealing the deal in my younger days.

During the comparative anatomy studies at Yale, I had the opportunity to dissect a number of harbor seals that had been found dead along the New England coast. Several of the animals had heartworm. The parasite, although similar in appearance to that found in dogs, was a different species. We knew that the dog parasite was transmitted by mosquitoes, but the insect vector responsible for the seal heartworm was unknown.

At Washington State, my research focused on finding that vector. As it turned out, the northern fur seal found on the Pribilof Islands in the Bering Sea, off the coast of Alaska, had a similar species of parasite, the difference being that it was found beneath the seal's skin. Both the harbor seal of New England and the northern fur seal of the Pribilofs were also infected with lice. If I could prove the lice vectored the parasite in the fur seal, I could get funding to look at the lice of the harbor seal.

I spent two summers and part of one winter on this project, on St. Paul Island in the Pribilofs. During that time, I lived among the nearly five hundred Aleuts who populated the island. This was my first experience with immersion into a totally different culture, and it felt very natural. I didn't take the position that "I knew everything." Instead, I let Aleuts like Alfe Hanson teach me not only about the seals, but also about the arctic fox found on the island. In fact, Alfe became a coauthor of a paper I published regarding rickets in the St.

Paul Island fox population. As a scientist, I learned a lot, but as a person who felt neglected by the general population as a child, I felt somehow connected with these people who accepted me without qualification. I treated them with respect, and in return they treated me as an equal. This became one of those life-defining moments. The research went very well, but to complete the project I would have had to raise seals in Washington and then infect them, a process that wasn't really feasible.

I returned to Yale in 1978 as supervisor of the Orthopedic Research Laboratory. Along with Dr. John Ogden, chairman of the Orthopedic Department, and Alden Mead, we formed the Marine Mammal Stranding Center, which operated for two years. In the summer of 1981, I was a National Oceanic and Atmospheric Administration (NOAA) observer aboard a Japanese salmon factory ship in the North Pacific. An unfortunate consequence of the fishing technique was that Dall's porpoises often got entangled in the nets. My job was to perform the necropsy to determine their health status and relative age. Mostly, I was looking at the parasites, but I also cut the flippers off and brought them back with me to Connecticut. At the Orthopedic Lab, we did a study on how to determine the age of the porpoises by x-raying their flippers and comparing the degree of skeletal development to the development of the reproductive organs noted during the necropsy. In the summer of 1982, I worked in Newfoundland looking at whales that get caught in codfish nets.

When I returned from Newfoundland, I went back to Washington State to do a dissertation on the biomechanics of seal swimming, which involved using X-rays to document the development of the seal's flipper and how it relates to biomechanics. I stayed there two years, living on $10,000 a year, when an offer came from Thomas Jefferson University in Philadelphia to teach in their X-ray program. I gave up graduate school for the job at Jeff, my first teaching job in imaging, which paid around $27,000.

While at Jefferson, I met Gretchen Warden, who had just been named director of the Mütter Museum. The Mütter was the country's last museum of pathology, an incredible resource. The museum had in its collection, since 1875, a unique specimen everyone called the Soap Lady. Her body was encrusted in this soap-like material termed *adipocere*. I had originally read about Gretchen and the Soap Lady in *The Philadelphia Inquirer*. The story focused on Gretchen's desire to understand the extent of deterioration in the Soap Lady and what might be required to conserve her. The museum had approached the University of Pennsylvania, which felt the mummy would have to be moved to the school for this type of investigation. Gretchen believed, quite correctly, that the Soap Lady was too fragile to be moved.

When I read this, I contacted Gretchen and proposed that I bring a portable X-ray unit and a team of students to the Mütter. I described some X-ray techniques I had used while working in emergency rooms, where patients were in too much pain to be manipulated into traditional positions, and she was intrigued enough to say, "Okay, let's see what you can do."

This was fantastic, except for one thing: I didn't have a portable X-ray unit. Fortunately, Philadelphia was the right place to be. It has some of the nation's top medical schools, podiatry schools, and osteopathic schools. I was able to locate a funky old portable unit owned by a podiatrist who was a recent grad. When I went to pick it up, I noticed there was no timer on it. I asked the guy, "What's your average exposure time for a foot?"

"Seven bananas," he answered. Seeing that I had no idea what he was talking about, he further enlightened me. "Yeah, you know—one banana, two banana, three banana . . ."

When Gretchen saw the unit, she was kind of amused at first, but she was very pleased with the results. I brought two students, Frank Cerrone and Mike Schenk. We x-rayed the Soap Lady over a period of several weeks, and eventually determined that she had not died, as legend had it, during the cholera epidemic that hit Philadelphia in the late 1700s. We found pins and buttons embedded within the soap-like material around her body. The pins had rounded heads, and the buttons had four punched holes in the centers. Neither item was manufactured before the early 1800s.

There is a permanent exhibit now at the Mütter on the Soap Lady project. It includes a photograph of me, my two students, and Gretchen, which has great meaning to me, as Gretchen passed away in August of 2004.

By this time, I had developed a philosophy on life. There seemed to be real value in moving every four years or so. New surroundings, new faces, and new challenges helped me avoid getting soft and comfortable. If you go to a new place, you have to work really hard, take some chances, do new things, and respond to a variety of personal and professional challenges. If you don't, how can you ever know what your true capabilities are? I think people who make meticulous plans miss out on a lot of what life is about.

> **I think people who make meticulous plans miss out on a lot of what life is about.**

I received an offer from Gulf Coast Community College to set up an X-ray program, so in 1988—four years after arriving in Philadelphia—I moved south to Panama City, Florida. Four years after that, in 1992, I returned to my old stomping grounds at Quinnipiac University, where I am now a tenured professor in the College of Health Sciences. Ron Beckett, who would become my friend and later my co-star in *Mummy Road Show*, was the man who hired me.

It's kind of scary to think that was more than ten years ago. I might be comfortable at Quinnipiac, but I'm not bored. I'm not sure I believe in destiny, that the events in my life led me to working with mummies. But I know this: *Mummy Road Show* was a wonderful experience. I don't think it would have worked with just Ron or me. It needed both of us playing off each other. Some shows had a eureka moment, and others didn't. It was truly reality TV.

We (almost) never let anything come between us on *Mummy Road Show*.

Chapter Two

We See Dead People—
Ron Beckett

You can tell an awful lot about how a mummy lived and died if you know what to look for, and, most importantly, how to look for it. Jerry and I had been combining our capabilities in radiology and endoscopy to unravel mummy mysteries for a couple of years when we and Engel Brothers Media came up with the concept for a one-time, one-hour show on the non-destructive analysis of mummies, and the idea grew into *Mummy Road Show*. The series appealed to us on a number of levels, but one of the really exciting things about it was the fact that our techniques would evolve at an accelerated pace.

Jerry and I strut our stuff *Mummy Road Show*–style in Guanahuato, Mexico.

The X-ray machine and the endoscope are devices familiar to most people. One "sees through" an object while the other looks around inside. Each instrument has its own advantages and disadvantages, but to my mind they are ideally paired for the non-destructive investigation of mummies.

I use an endoscope on a plastinated heart—more on plastination later.

Endoscopy is my area of expertise. For those of you who don't know, an endoscope is a tiny camera attached to the end of a flexible rod, which is connected to a monitor and recording device. It was first used in the study of mummies in the early 1980s, though the groundwork was actually laid a decade earlier. Advancements in fiber-optic technology in the 1970s offered amazing flexibility to the endoscope. A fine-wire remote control enabled the operator to maneuver it into body cavities and organs. And with the viewing area lighted, it was possible to see things that were once completely inaccessible. Video endoscopes were developed in the 1990s. They provided even greater resolution, which made interpreting data easier and more effective.

Mummies for Dummies

What is a mummy? Simply put, the preserved remains of a dead person. The defining factor is that the soft tissue must still be there. In other words, a mummy is skin *and* bones (although in rare cases, such as mummies created by peat bogs, it is possible to just have skin). Mummies can be formed by a number of natural processes, like the famous Ice Man in the Italian Alps, and, of course, by artificial means, like many Egyptian mummies. Although most of the mummies that are being studied are very old, technically age is not a factor. There are mummies out there that are three thousand years old, three hundred years old, and thirty years old.

Mummies exist in a lot of cultures. Individuals preserved prior to burial run the gamut from kings and queens to the homeless. Some were entombed in elaborate structures, such as the Egyptian pharaohs (or Vladimir Lenin, whose mummified remains are held in a climate-controlled vault in Russia). Some never made it into the ground, as was the case with the mummies on display at carnival sideshows during the nineteenth and twentieth centuries. Mummies created by nature are often happy accidents, produced by a unique set of circumstances that are very difficult to fully explain. Indeed, there are many natural mummies that have yet to yield their secrets. What all mummies have in common today is that they are objects of fascination, reverence, and scientific curiosity.

The cartonage of a mummy we investigated at the Chatham-Kent Museum in Ontario for "An Egyptian Souvenir."

Mummified remains are not limited to humans. Over the centuries, people have gone to great lengths to preserve animals, too. Some creatures held great symbolism and were welcome in ancient tombs, while others were probably household pets. Although the study of mummies is almost completely focused on humans, scientists do get excited when they encounter mummified animals because of their scarcity and cultural significance.

The mummies with which people are most familiar come from Egypt, where a large and complex civilization flourished for many centuries, dating back between two thousand and six thousand years. Like the people of many cultures, the Egyptians believed that death was a doorway to the next life, the afterlife. Their belief system, however, held that a person would need his or her body to be intact at the next stop. Thus, corpses were prepared and preserved, the organs removed, and the body treated with a number of substances—including preservative resins—and in the final stage, wrapped in linen.

At first, Egyptians buried their dead in the sand, much as bodies are interred in cemeteries today. Individuals of high status, however, began demanding that they be buried in more dignified surroundings, where they would also have room for the many possessions they hoped to bring with them into the afterlife. This eventually led to the magnificent tombs and other structures, including the pyramids, which would house the bodies of the pharaohs and other Egyptian royalty.

Not surprisingly, the business of preparing the dead assumed an important and mystical place in Egyptian society. The science of mummification reached high art, and even had its own god, Anubis, who is commonly depicted as a jackal (or a man dressed as a jackal). During mummification itself, a priest would wear a jackal mask as he presided over the ceremony. The mummification process could take up to two months.

In recent years, the public has become more aware of the mummies produced by a number of ancient South American cultures, particularly in the arid regions of Peru and

Chest cartonage from an Egyptian mummy.

Chile. At first glance, the burial rituals of these people may appear modest compared to those of the Egyptians, as the bodies were often bent into a sitting position and wrapped in fabric. Some of these mummy "bundles" contain fantastic textiles and artifacts, and can weigh over 150 pounds. As the study of South American mummies has advanced, however, we know that their burials were no less

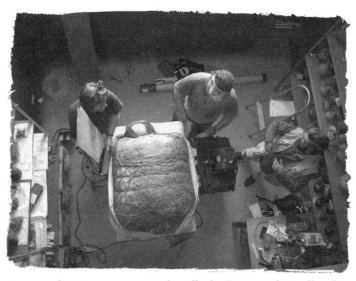

Jerry and I X-ray a mummy bundle in "House of Bundles."

elaborate, and held as much, or even more, cultural significance.

The word *mummy* itself comes from the Persian word for bitumen, "mummia." Bitumen is a mineral formed from tar (also known as pitch), and was highly prized for its restorative powers until relatively recently. How preserved corpses came to be called mummies probably goes back to the height of the mummia trade in the Mediterranean, during the Middle Ages. Travelers to Egypt may have mistaken the dark surface of mummies for a coating of bitumen, and somewhere along the way the entire body was referred to as "mummia."

The Egyptians, all too happy to exploit this new "source" of bitumen, pulled mummies from their tombs, ground up the entire corpses, and sold the powder to pharmacists and tourists alike. When they ran out of mummies, they began tanning fresh bodies and grinding them up, too. Although Persian mummia was always the most highly prized, Egyptian mummia was far more plentiful and thus in much wider use. In the 1500s, it was part of every druggist's inventory, from the Middle East to Europe. In the 1600s, the difference between the Egyptian powder and Persian powder was better understood, and mummia quickly fell out of favor. The name stuck around, however, and soon evolved into its modern and more familiar meaning.

For me, the endoscope's maneuverability is one of its greatest assets. When we were in Popoli for the *Mummy Road Show* episode, "Tales from an Italian Crypt," I used the endoscope to spot and remove a large kidney stone from a mummy. Later, we found calcifications in a tissue sample from this same mummy that possibly indicated tuberculosis, a significant discovery.

That's me in the hole, maneuvering my "tomb cam" into position.

"Mummy Rescue" provides another good example of the endoscope's versatility. In this episode, an earthquake in a large desert region in southern Peru exposed a large number of mummy tombs we wanted to investigate. We needed to get a view of what was inside, so I fashioned an endoscope into what we called the "tomb cam." This was helpful because we were able to see what impact the earthquake had had on the tomb contents, how the bodies had been placed in the tomb, and what items had been buried with them. In turn, our team was able to plan the excavation more effectively. (There's a lot more on both of these stories—a fascinating revelation and an unforgettable experience—later in the book.) This kind of information—the special relationship of grave goods—is often essential to piecing together a mummy's story.

Sonia Gúillen with a Chiribaya mummy.

Hey, Taxi!

Maybe the biggest challenge in "Mummy Rescue" was finding a power source for the tomb cam. Here we were, in the middle of nowhere, and we needed a generator. Then, as if on cue, a dilapidated taxi came by crowded with a Peruvian family. We flagged it down and convinced the driver to let us plug a power converter that I had with me into the taxi's cigarette lighter. We were all very excited, until the scent of what smelled like an electrical fire began to rise around us. This was one of the few times during *Mummy Road Show* that one of our improvisations didn't work. I looked at Jerry, and did the natural thing—I started laughing uncontrollably. The taxi driver didn't see the humor in the situation. Afraid his cab was about to blow, he fled on foot, with his passengers right behind him. After assuring them that the taxi was not a ticking time bomb, we gave the driver a few *soles* (Peruvian currency), and off they went. We drove back to the research facility, got a gasoline-powered generator, and returned with a more reliable power source for the tomb cam.

Jerry's experience as an X-ray technician—he's been in the field for more than three decades—was incredibly valuable. At least fifteen of those years were spent in the emergency room. Oftentimes, trauma victims can't be moved, which forces the X-ray technologist to take a creative approach to obtaining the images he needs. This gave Jerry great insight into working on mummies. He was always excited by the challenge of a tough X-ray job. Jerry was like a veteran photographer; he became very skilled at manipulating exposure factors, using various types of film, and operating all sorts of X-ray equipment, particularly mobile units.

Jerry and I take a break with the Engel Brothers Media crew.

What can be determined from X-rays? Age, for one thing, if the mummified person died sometime before their twenty-fifth birthday. Jerry figures out age primarily by looking at the bone development of the arms and legs. On people mummified after the age of twenty-five, however, this method isn't nearly as precise, so you have to focus on the degenerative changes that occur with age and physical stress.

Teeth are also good indicators of age. From birth until the teen years, teeth appear in a specific sequence at predictable ages. Once all the adult teeth have appeared, relative ages can be approximated based on wear and tear of the teeth.

In many episodes of *Mummy Road Show*—like "Mystery in a Bundle" and "House of Bundles"— we needed to ascertain the sex of a mummy. X-rays of the skull and pelvis were helpful in these cases.

Jerry sets up his X-ray equipment during "Luck of the Mummies." Fortunately, he is not claustrophobic.

Abnormalities in a skeleton—anything from a traumatic injury to degenerative and pathologic changes—are always obvious on X-rays. Something we always look for is commonly termed "Harris" growth arrest lines. We saw

these lines on the bones that formed the legs of Princess Anna, in the episode called "Princess Baby." Anna was a mummified child (the daughter of King Ludwig, a fourteenth-century Bavarian emperor) whose body was lovingly cared for in the monastery in Kastl, Germany. The parallel Harris lines on her X-rays indicated that a process, maybe an infection, stopped her bone development for a period of time. Since there were several of these lines, it

Jerry and I peruse the CT scans of Andy the carnival mummy.

Without the right science, determining a mummy's age becomes a highly subjective endeavor. These calcium deposits on the arm of a mummified nun led to the story that she was more than 100 years old when she died.

told us that Princess Anna had suffered from health issues more than once in her short life.

Sometimes, if the internal organs of a mummy are preserved and in good condition, X-rays reveal fascinating things. In "Homemade Mummies," we studied two mummies, a pair of inmates from the West Virginia Hospital for the Insane, who were prepared in the late 1800s by a farmer and amateur scientist named Graham Hamrick using a process he had patented. This episode was a lot of fun. Jerry x-rayed both mummies in a museum that was once the train station in Philippi, West Virginia. In one of the mummies, he detected what appeared to be a lesion in the left lung

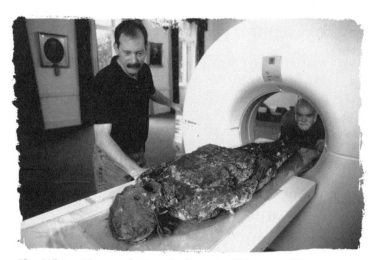

The Mütter Museum's Soap Lady gets a CT scan with technologist David Lindisch.

that was the result of tuberculosis. TB was a major health problem in the United States in the latter half of the nineteenth century and early in the twentieth century. In fact, it was one of the primary causes of death during that period. That's why Jerry's finding was so meaningful. Here we had a rare case, a perfectly preserved individual that served as a time capsule, and the X-rays offered important clues about how she died.

One of the shortcomings of a conventional X-ray is that it produces a two-dimensional image. This is where Computed Tomography, or CT, comes into play. Jerry likes to put it this way: Think of a sliced loaf of raisin bread. If he takes an X-ray of an unwrapped loaf from the side, he can't tell the exact location or depth of the raisins inside. If he opens the package and separates the slices, the exact position of the raisins is clear. CT works the same way. Using X-rays to circle a mummy, Jerry is able to digitally collect data and produce axial images, or slices. The thinner the slice, the more precise the data.

CT images provide much higher resolution than standard X-rays, and in turn yield more information. For example, Jerry can stack the slices and create a three-dimensional image that shows great detail. This allows us to see what lies beneath a bundle without unwrapping it.

As scientists, we know that an autopsy would allow us to determine with nearly 100 percent certainty what is going on inside a mummy. But we would hate to de-

I do my thing with a mummy bundle from Cajamarquilla, Peru.

stroy anything of cultural significance. We prefer our non-destructive approach because it maintains a measure of respect for the dead, and because the mummies are preserved for later study.

In the case of South American mummies, which are typically found bundled in textiles, the cultural significance may be how the bundles are constructed, or where certain objects are placed inside the bundle. (The term "cultural significance" refers to a variety of things, such as an individual's status, role or occupation in society, funerary and burial practices that may say something about belief systems, and how people cared for one another.) Once you unwrap a mummy bundle and take it apart, that's that—it is apart, and you've lost all that information. More specifically, you've lost the special relationship of artifacts within the bundle. Objects may have been placed in a particular location or manner, which could reveal something important about the culture. What we *don't* know about mummies today could fill many books, so there is a desire on our part to preserve information that might be easier for archaeologists or anthropologists at some later date to interpret. We can do this through our imaging techniques.

Using X-rays, you can begin to determine patterns. Are you looking at a typical mummy bundle, or is this an unusual one? If one bundle is different from, say, one hundred others, what does this tell us? What else should we be looking for? This is an exciting prospect for a scientist, but for us on *Mummy Road Show*, it was also a source of great frustration. We wanted to stay in these places and check out all one hundred bundles. But sometimes the schedule only allowed us to examine a few. It was really difficult to leave in those situations.

Monkey Business

X-raying a mummy is usually the quickest way to determine whether it is genuine or not. The people who made and sold fakes generations ago knew their owners would never unwrap them, and I doubt they ever imagined that their counterfeits would be subjected to something like an X-ray. Even when you find a bogus mummy, it's thrilling nonetheless.

During the filming of "Egypt, California Style," we were at the Rosicrucian Egyptian Museum in San José, looking at a mummy the museum had purchased from a Neiman Marcus Christmas catalog (more on that later). While I was endoscoping the mummy, Jerry went exploring in the museum, which featured quite a few animal mummies. Included in this "menagerie" was what everyone believed to be a mummified baboon, which was on display in a glass case.

Jerry was instantly intrigued when he spotted the baboon, and decided he was going to X-ray it, which he could do right through the case by taping Polaroid film behind the baboon. This is a very tricky procedure, almost like an X-ray art form. Not surprisingly, this is an area where Jerry really excels. In fact, he may be the only radiographer who employs this technique. Instant radiography is done like a standard X-ray procedure, except there's no darkroom needed. The images—presented on Polaroid film, which develops in a minute—are of very high quality.

I could see Jerry piling up boxes to get the portable machine to the right height. That always worries me, because of how fragile the unit is. But he hasn't lost one yet. Anyway, the Rosicrucian is a wonderful place, and the curator there, Lisa Schwappach-Shirriff, is a brilliant woman, with such energy and enthusiasm. When she heard we were going to x-ray her prized baboon, she got really excited.

After I completed my scoping, everyone went over to see how Jerry was doing. When he finished the exposure, we processed each 8" x 12" film packet and assembled a dozen or so of these pictures into a baboon mosaic. Our jaws just dropped when we saw the whole picture. There was no baboon at all—it was simply a vase skillfully made to *look* like a baboon! And we

The Rosicrucian's prize mummified baboon, before . . .

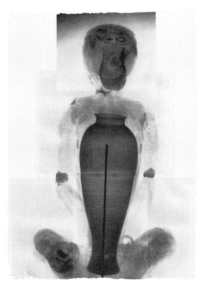

. . . and after Jerry's x-rays.

weren't even sure if it was a 2,000-year-old vase, as the museum had believed. It could have been from the 1920s, something a slick mummy vendor had put together and sold as a rare, sacred baboon.

I have to say, Lisa took it well. She thought it was kind of cool to have a fake "baboon." The good news is that we went back and checked out other specimens in their mummy menagerie and made some other significant finds.

As I mentioned, my endoscope comes in handy when we want to get an up-close look at something that either shows up in the X-rays, or if we just want to do some exploring. It requires a small opening, which is usually not a problem with mummies due to deterioration, and it gives you a nice view of the interior on a video monitor. Again, an endoscope pairs so well with an X-ray in these situations, because each generates visual information the other cannot.

For example, in "Carnival Mummy," we traveled to the Ripley's Believe It Or Not! Museum (which also had a huge warehouse) in Orlando to study a carnival mummy named Andy that was going to be put on display at the Ripley's Museum in New Orleans. Edward Meyers, who was in charge of exhibits and archives for every Ripley's museum worldwide, wanted to know everything we could find out about Andy. There were two wildly different legends surrounding him. One said he was a carnival worker whose dying request was to have his body mummified and then featured as a sideshow oddity. The other said he was thousands of years old, his body preserved naturally by desert sands. In this episode, it was a combination of X-rays and endoscopy, plus CT scans, that cracked the case.

We have investigated a lot of carnival mummies in the U.S. They were popular attractions for many, many years in traveling sideshows. In fact, many of the "mummies" that toured with sideshows in the nineteenth and twentieth centuries were actually manufactured for the sideshow industry—

in other words, they were fakes. Those that were real were probably bodies of outlaws or homeless people that were preserved by embalmers using arsenic, but never claimed, and for one reason or another never buried. Later, they were sold to the traveling carnivals.

Our initial examination of Andy revealed several interesting insights. He had incisions in his neck, abdomen and upper legs, all of which were closed with "baseball" stitching (so named because of its similarity to the stitches on a baseball). We took some digital photos and emailed them to our friend, Ronn Wade, at the University of Maryland School of Medicine. Based on the stitching, Ronn concluded that Andy died some time around 1920. This finding immediately refuted legend #2.

Next, we got to work on legend #1. Jerry's x-rays were fantastic. The first thing we saw was what looked like a nail embedded in Andy's posterior nasal cavity. (That's right, a nail in his head!) This finding gave credence to the story of Andy being a carnival worker. Maybe he had been a human block-head—a person who drives nails and other objects up his nose. But something was strange here. The nail was very deep in the nasal cavity and at an odd angle. Was Andy the victim of a block-head trick gone wrong?

To find this out, I snaked my en-doscope up into the nasal cavity. I could see that the tip of the nail did not penetrate the back side of the nasal cavity. This told us that the nail was not the cause of death. In fact, it was most likely placed in the nose *after* death. We wondered if this was perhaps done as a final tribute to Andy and his career as a blockhead.

If you know anything about carni-val people—and we do—this was a neat find. They are among the most

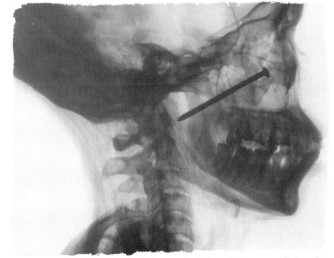

An X-ray revealed a nail in Andy's head: Cause of death or loving tribute to a fallen comrade?

The amazing Molotov shows his stuff. This is how it always starts—a few minutes later, we were the ones standing against the board!

fascinating people in our culture today. These individuals are very dedicated to their craft, and tend to be a really close-knit group. In Andy's case, it would be well within the realm of possibility that this was his "favorite" nail, and that he made a deathbed request to have it buried with him in this manner. We have a carnival friend named Harley Newman who does a similar act, using everything from spiked heels to a Black & Decker drill. We wouldn't be surprised to find something unique embedded in his nose when he goes.

But that wasn't all we learned about Andy. On the first X-ray Jerry shot, it appeared he had suffered a broken sternum (or breast bone). But our initial observation was later proved wrong. Without an autopsy, determining the cause of death is very elusive—unless it was a traumatic demise that left signs on the skeleton. On Andy's CT, taken by Dr. George Stanley at the nearby Winter Park Diagnostic Center, the story literally unfolded layer by layer. Because his heart and lungs were so well preserved (thanks to the arsenic embalming process), a picture began to unfold. Andy died from a condition called tension pneumothorax, which was caused by his

Jerry, Willy Cock, and I examine the X-rays of a mummy bundle.

broken ribs, not a broken sternum. His rib fracture penetrated the chest cavity, allowing air into the space outside of the lungs. The resulting air pressure collapsed one of his lungs and moved other internal structures like the heart around, which put pressure on the major blood vessels in the chest. These combined effects of a tension pneumothorax can lead to death if not treated immediately, and apparently in Andy's case, medical help never arrived. This realization came to Jerry and I almost simultaneously as we were looking at the CT scans. It was a remarkable moment for us. It's rare when you can say with at least 90 percent certainty how a person may have met his end.

The techniques we used during the show are very relevant to what is happening today in the examination of Egyptian mummies. Recently, attention has shifted from a focus on royalty toward the study of more "ordinary" mummies and their tombs. We used to draw our assumptions about Egyptian life from the royal tombs—from the paintings, the artifacts, and the mummies themselves. But we now know this just doesn't provide the whole picture. We can learn a lot more about the life of the middle-class Egyptian from middle-class mummies, and the life of commoners from mummies of commoners. Using X-rays and endoscopes, we can also get a clearer picture of what diseases they were fighting, what injuries they suffered, and the most common causes of death for various ages and social strata.

Arsenic and Old Face

The best "modern" mummies we see are the ones embalmed with arsenic solution. Mummification is the preserved result of embalming, but embalming does not always lead to mummification. With just the right degree of embalming, someone may only be preserved for three or four days (hopefully until after the funeral). With a different mix, they can last for centuries. Arsenic embalming gives the corpse almost a wooden look, and preserves everything beautifully. It also makes bodies very dense and very firm, making them easier to transport. In the sideshow business, this actually made them more marketable than Egyptian mummies, which would start turning to dust when you moved them from town to town by train or truck.

Arsenic embalming was first practiced in France in the 1840s. Every embalmer had his own home brew, but as a rule they used up to ten pounds of the stuff per body. It was outlawed in the early 1900s, because of the danger the chemicals posed to the living. In turn, it became hard for embalmers to lay their hands on it. I've been told that it was still in use by some old-timers in the 1950s and '60s, and if you talk to embalmers today, they still say arsenic is the ideal way to embalm. We would tend to agree. When you come across a mummy preserved in this way, you get a really good sense of what they looked like at their time of death. Their internal organs, rarely intact in other mummies, are easier to examine as well.

The reason arsenic was outlawed is because it was making the people who worked with it sick. It's really nasty stuff. In fact, now we are starting to see issues with Civil War cemeteries, where the wooden coffins have dissolved and the rain is soaking down through these densely packed, arsenic-laden bodies. There are already cemeteries where the groundwater is contaminated in the surrounding area. This is going to become a bigger and bigger issue as time goes by.

Once in a blue moon, we are able to autopsy a mummy. Think about how rare this is. Why would someone who owns or cares for a mummy let you do that? That's a good question, because an autopsy is obviously an invasive procedure, which happens to run counter to our normal approach to mummy research. But an autopsy can tell you things about how a person died that we simply can't uncover through X-rays and endoscopy. Sometimes, this is the information that a mummy's caretaker wants to know above all else. This was the opportunity presented to us in "An Unwanted Mummy," an episode that sought to uncover what really happened to Hazel Farris, one of our all-time favorite mummies.

We were working against the clock as soon as we arrived in Nashville to study Hazel. She had been a sideshow attraction for decades, after showman and entrepreneur Orlando C. Brooks purchased her in 1907. Her current owner—a caring woman named Teresa, who was a great niece to Brooks—planned to have her cremated. She felt Hazel had been exploited for too long, and it was time for her to rest in peace. This was a noble sentiment, and it also made it easier to convince Teresa to let us conduct an autopsy. In the end, we all wanted what was best for Hazel.

We were very respectful in our request, and assured Teresa that we were merely after the truth about Hazel and the stories surrounding her life and death. Without an autopsy, we explained, we could not separate fact from fiction and give Hazel her due. We also explained that autopsies are done all the time to help determine how someone died, and reiterated that our objective was not to desecrate the body. Teresa agreed, and we got to work.

All mummies that are on display in sideshows have great stories behind them, but Hazel's history was more elaborate than most. The "true" story of her life supposedly went like this: She and her husband lived in Louisville, and both had drinking issues. On August 6, 1905, she shot and killed him while arguing over a hat, which touched off an unspeakable bloodbath. Three deputies were dispatched to apprehend Hazel, and she managed to shoot all of them. The sheriff followed, wrestled her to the ground, and in the struggle shot off her ring finger. But Hazel got the better of the sheriff and shot him dead, too, bringing the body count to five. With a hefty reward offered for her capture, Hazel fled to her hometown of Bessemer, Alabama, where she became a lady of the evening. She got into a relationship with a man and confided in him, recounting her past indiscretions. This man liked Hazel, but he found the $500 reward more appealing, and turned her in. When she saw the police coming down the street to arrest her, she supposedly committed suicide by drinking a mixture of whiskey and arsenic, a potion said to have magically preserved her. She was taken to a local furniture store, which made the coffins in town, but her body was never claimed. The store owner sold her to Brooks for $25, plus the cost of the coffin. She landed on the carnival circuit and spent decades as a sideshow mummy. The showmen even produced a WANTED poster for Hazel, with a woman's face on it.

Hazel was eventually passed down through the family members that owned her during her carnival years. Returned from the circuit, she was kept under a couch, and the kids would literally take her out and play with her, as if she were a big doll. The family cleaned Hazel up every so often—we found traces of Borax on her—but eventually she was passed down to Teresa, who stored her in a garage, where she started to get a little moldy. Eventually, Teresa decided it was time to lay Hazel to rest once and for all. Luckily, she was also curious to know whether Hazel's story was true. This is when we were called in.

We thought Hazel's legend was really intriguing, but there were a couple of key details that didn't add up. For starters, there were no newspaper accounts of the five slayings—in Louisville or anywhere else. The staff at Engel Brothers is always very meticulous in their research, and they could not find a thing about Hazel. Well, you don't shoot your husband and four cops dead without making headlines. The other red flag was the arsenic story. I've actually *had* whiskey and arsenic, and lived to tell the tale. (Just kidding—though I've spent a few nights drinking whiskey, and have indeed felt dead the next morning.) And even if Hazel had ingested pure arsenic, it might have killed her, but there was no way it would have preserved her. It takes an enormous quantity of arsenic, injected directly into the veins, to accomplish this. As for the other components of the story—we didn't even know if her name was really Hazel—we hoped to find out what we could in order to either support or refute them. And, of course, we hoped to produce a lot more information with the X-ray and endoscope.

Hazel's X-rays revealed some interesting details about her real life. There were some unusual shadows that showed up on her chest, but we were not sure what to make of them. Remember, while we weren't the first to study mummies with modern science, no one was teaching a course in interpreting mummy X-rays—we were learning as we went along. I got in there with the endoscope, but I couldn't say with much certainty what the shadows were. She was preserved so well that her internal organs impeded the endoscope.

Jerry was able to determine that she did indeed lose her finger while she was alive, which was great, because sometimes parts snap off mummies (or someone breaks off a body part for a souvenir) and their owners simply further embellish their stories. To him, it looked like a clean amputation. He even detected some bone growth after the finger was lost. Was it blown off by a sheriff's pistol? Probably not. We would have seen more evidence of shattering.

Another interesting thing we found was evidence in Hazel's pubic bones that suggested she had given birth during her lifetime. No children were mentioned in her sensational story, which again suggested that it was a fabrication. I think putting a kid into the mix would have made this mummy even *more* exciting to carnival goers. Be that as it may, this new piece of information placed Hazel in a whole new light. She could very well have descendants walking around today. For Teresa, it made her even more determined to lay Hazel to rest.

Meanwhile, our work revealed nothing that would suggest Hazel had died by her own hand. Even the CT scans presented us with more questions. The shadows on her lungs were the most likely cause of death, with tuberculosis a prime suspect. The problem was that, regardless of how well preserved Hazel was, and despite our best efforts, we could not say for sure what had killed her, or what these shadows were. Jerry and I knew that the only way we could unravel this mystery and tie everything together was to perform an autopsy on Hazel.

Larry Cartmell—a pathologist and mummy specialist from Ada, Oklahoma, of all places—performed the autopsy for us. He was great about moving along at a pace at which Jerry and I could ask him questions. Instead of just whipping through the autopsy in ten or fifteen minutes, he really took his time and made the most of our experience with him.

What an opportunity! Hazel was a fabulous mummy, remarkably well preserved—a gold mine of information. Every tissue and every cell had to have been impregnated with arsenic solution for her to be in the shape she was. Her organs were incredibly preserved. Before removing the chest wall, I used the endoscope to see between the chest and the lung. The entire lung was "stuck" to the inside of the chest wall, a condition known as a pleural adhesion, which is often caused by massive pneumonia. The autopsy confirmed the existence of a series of adhesions, which must have been very painful. Ultimately, however, it was the pneumonia that led to Hazel's death.

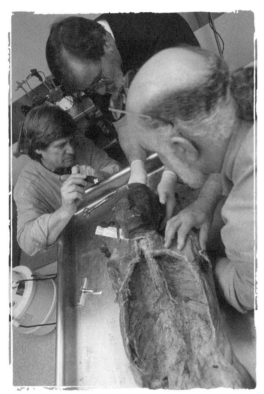

Jerry and I work with Larry Cartmell during Hazel's autopsy.

At one point, Larry removed the heart and lungs and let Jerry x-ray them. The solution that the people who had mummified Hazel had used made the blood clot in her lungs, which is why we were seeing shadows on the X-rays that appeared to be lesions. The clots were clearly postmortem, the result of some interaction between the arsenic and the blood. We could tell because when blood clots pre-mortem, it sticks to the walls of the blood vessel, and this was not the case.

Hazel Farris changed our lives. Being able to autopsy her demonstrated to us how far we had come in terms of combining our X-ray and endo techniques, and how far we had to go. It also got us excited about the prospect of keeping our show on the road, so to speak, and it made for an interesting one-hour special at the end of our first season. We had done thirteen shows already, plus the Hazel special. Time had flown by. Fortunately, the National Geographic Channels agreed to pick up the series for two more seasons, and *Mummy Road Show* lived on. For the next couple of years, we would be seeing a lot more dead people.

Chapter Three

Have Bones, Will Travel—Jerry Conlogue

Being full-time professors and appearing in thirteen episodes of *Mummy Road Show* that first year did not look possible on paper. Not that we would have let *that* stop us—but we knew that we would have to function as part of a very large team. Ron is an experienced performer and natural ham, and I'm not exactly shy, so we were not concerned about what would transpire in front of the camera. It was the other stuff that was scary. There is an unbe-

Ron does his thing during the filming of "Turkish Tomb Mummies."

lievable amount of work involved in producing a television show, from securing permission to work on a mummy, to the travel logistics of getting a film crew to the mummy, to getting *us* to the mummy, and, finally, ending up with enough usable footage to make a compelling and scientifically valid show.

Needless to say, it all worked out. The research staff and producers at Engel Brothers Media were incredible. We still talk about what a superb job

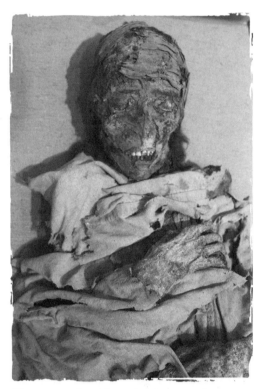

The research and production people on *Mummy Road Show* came up with some wonderful mummies during the three years we did the series. Here's a portrait of the "Blonde Nun" from Turkey.

they did. These folks would beat the bushes for mummies, and let us know when they found "a good one." Sometimes they would ask us what we thought, and sometimes they would just tell us, "This is the one." They aimed high and lined up a lot of mummies that no one had studied before. Part of the appeal, of course, was that the caretakers of these mummies would get exposure that would benefit them in one way or another. Even so, Ron and I always approached them as fellow scientists. (And, indeed, some were.) And if there were other scientists involved on the show, they always got copies of the X-rays and all the data we gathered so they could continue their work after we left.

What opened a lot of doors, I think, was the fact that National Geographic was behind the show. That name has built up so much respect over such a long period of time. National Geographic has a rich tradition of really decent science and a reputation for presenting things in a responsible and interesting way.

Over the course of the series, the episodes developed in one of three ways: Sometimes Engel Brothers would contact us with mummies they had found or stories they thought were worth pursuing, and we would get involved in whatever way made sense. Other times, people would communicate directly with one of us or Engel Brothers, and we would take it from there. And of course, we knew where some great mummies were already.

The producers at Engel Brothers would function as go-betweens to set up the shows, but as thorough as they were, Ron and I sometimes found we still had to seal the deal when we arrived. On the surface, the primary goal of *Mummy Road Show* was to examine mummies and accumulate scientific data that might tell a story. But there was much more to each episode. We were studying the remains of someone who had once walked this earth. Dealing with the caretaker of those remains was a sensitive issue.

Until you have worked with the living in a medical environment and seen death firsthand, it is difficult to fully grasp the responsibility you must accept before proceeding with the task of mummy study. Before you begin an examination, it is important to convince the caretaker that you have sincere intentions, that you will treat the remains with respect and not damage them. Protecting the mummy always takes precedence over a scientific investigation.

The key in many cases came down to demonstrating some cultural sensitivity. We knew how important it was to be accepted by the people we were working with, which meant showing respect for their customs and beliefs. We were very aware of the "Ugly American" image, and did whatever we could to dispel it. I mean, you cannot sit down to a meal with someone and say, "Yuck, I don't want to eat that." In some instances, your hosts are probably offering most of the food they have, which means they're going to go without.

I remember once going to Leymebamba, which is a remote town deep in the Peruvian Andes. For this trip, I took Barry Anderson of Ripley's Believe It or Not! Museum in Orlando with me. Poor Barry had never really been out of the country. We had this seven-hour flight from Miami to Lima, and then we took another two-hour flight from Lima to Chiclayo. We got picked up by a vehicle in Chiclayo, and what was supposed to be a twelve-hour ride in this little minivan turned out to be closer to eighteen hours. When we started the ascent into the Andes, on a winding road without any guardrails, Barry was totally freaked out. Once we got to Leymebamba, we were shown to our room. Fortunately, for Barry's sake, we weren't handed a chamber pot. There was actually a bathroom. Still, Barry was really worried. There were spiders in the room, and he thought for sure he was going to come down with some disease.

It was festival week in Leymebamba, and during this week it's not unusual for people to invite you into their homes, which

We were studying the remains of someone who had once walked this earth. Dealing with the caretaker of those remains was a sensitive issue.

were like nothing Barry had ever seen before. We went to one home that had a dirt floor and very quaint, but very basic furniture. But they pulled out all the stops and cooked whatever they had to honor us. We weren't sure what it was, and I could see Barry was a little hesitant. I said, "As long as you take a bite, it will be fine. They don't expect you to clean your plate, but just make that attempt. Otherwise, even though they may not say it, you're really offending them." Being the trooper that he was, Barry took a bite, and he was amazed that he didn't get sick. In the end, he survived the whole trip—though I think he was really happy to get back to Orlando. It was definitely a trip he'll remember.

There were instances when we did get sick in foreign countries, but that can happen any time. For the "Blonde Nun" episode, we traveled to an historic area of Turkey called Cappadocia. It was beautiful. The landscape was dotted with all these striking rock formations. Even our hotel was carved into the rocks.

We were fortunate because two members of the Engel Brothers production team happened to be Turkish. Thanks to them, we didn't feel out of place. But our luck soon ran out.

I'm pretty sure it was the non-bottled water we had with lunch. One by one, the crew got hit by something nasty. Ron was the first to get sick. The car ride from the hotel to the museum was about an hour, and he didn't look all that great on the way there. It got worse at the museum. Ron was laid out on the floor in pain, even as we were filming. Then it became a domino effect. I got sick, the crew got sick. One of the production assis-

Cappadocia, Turkey. Enjoy the awe-inspiring landscape. Bring your own water.

tants from Engel Brothers actually passed out during a shoot one day. We were filming a scene, and in the background we all heard a loud *thud!* It was Sirin hitting the deck.

The bottom line is that whenever our hosts say let's go and do, you go and do. That was the case when we were in Amasya, a Turkish city not far from the Black Sea, for "Mummies from a Turkish Tomb." No matter how we were feeling, we had tea with the mayor everyday. It was sort of amusing. We didn't speak Turkish, and the mayor didn't speak English, so at most of our "tea parties" we grinned incessantly and never said a word. One of the production assistants, Aysin Karaduman, was Turkish. The mayor engaged her in nonstop conversation, which she would then translate for us.

Immersing yourself in a particular culture is important to do from a professional standpoint as well as a personal one. For instance, I had no idea what Buddhism was all about until we went to Thailand and spent time with the people in the monastery that housed Luang Pho Dang, the subject of our "Mummy in Shades" episode. His family

Ron was laid out on the floor in pain, even as we were filming. Then it became a domino effect. I got sick, the crew got sick. One of the production assistants from Engel Brothers actually passed out during a shoot one day.

claimed that he had once gone fifteen days without food and water. Initially, I thought, *Yeah, right. This can't be possible. No one is going to be able to go fifteen days without food or water.* But when you go there and immerse yourself in the Buddhist culture and philosophy, you begin to see how this could be true.

Ironically, some of our viewers may have thought we were being culturally insensitive during this episode, because the mummified Buddhist monk was wearing a pair of sunglasses. The truth is that the monks from the monastery had put the shades on him. They felt what was left of his eyes may have been a bit too much for visitors to handle. Also, it may have appeared that Ron was being disrespectful during one of the shots by not getting into the lotus position. But he had a bad hip, a condition that our viewers didn't know about until we incorporated his hip replacement into a subsequent show.

The entire Thailand experience was insightful, partly because our host, Phra Aduna, spent a lot of time with the two of us while we were at the monastery. While walking with him one day, Ron accidentally stepped on a snail—*crunch*—and Ron started apologizing profusely. Phra Aduna told Ron it was okay, that the snail had chosen his foot as the instrument that would help him rise to another level of existence.

On several shows, there was little time to convince people that we were interested in their religion or had studied their culture. In those instances, we were being sized up, and we were acutely aware of this. One thing that helped in this respect is that Ron and I drank whatever people offered us. Some of that stuff could fuel an aircraft, but we drank it until they were satisfied. Or until we experienced an emetic event. (In other words, until we puked.)

When we did "Mummy on a Mission" in Ecuador, we filmed a reenactment in the middle of a farmer's field. The landowner—who provided the horses and played the role of a Spanish soldier from the 1500s—felt a celebration was in order before we left the estate. He led us to a small building on the edge of the field, where there was a bar complete with a karaoke machine. The drink of choice was some home-brewed *aquadiente*. This stuff was potent. Tradition dictated that we drink from a ram's horn that was sitting on the bar. The horn held about three ounces, and after two horns' worth, you could have had us spinning around a pole minus our clothes. I recall being dressed as monks, and doing a duet with Ron. Unfortunately, it did not make it into the final cut of the episode. We were surprised that anyone was actually sober enough to film us!

Speaking of overdoing it, during the same episode, we received a grand welcome from a group of twenty men dressed in their finest suits. A celebration followed, and the curator of the museum really got into the spirit of the evening. He paid for it the next day. In fact, he showed up at the museum two hours late!

Another way we ingratiated ourselves with our hosts, particularly on our trips to Peru, was to show how much we cared about kids. We're both fathers, so this is natural for both of us. As you might imagine, people like it when you're really nice to their kids. But we never treated the children in a patronizing way. We love kids, and whenever possible, we would involve them in what we were doing. One of the constructive things to come out of this is that parents see you in a more positive light.

Ron scopes the ultimate "student body" at Naperville Central High.

In "High School Mummy," we visited Naperville Central High School just outside of Chicago, where they had a mummy they called Butch. (Actually, Butchina was a more appropriate name, but we'll leave this story for later.) This school was an incredible place. It was really clean, and the amount of money the town invested in the school was mind-boggling.

The kids at Naperville Central were most amazing of all. This is a great example of what happens when a community really cares about its school system. The students were phenomenal. They took such great care of Butch. At some schools, the mummy may have been turned into the mascot, or kids would have been constantly trying to steal it—but not here. Their concern for Butch was incredible.

This helped us form a real bond with the kids in Naperville. They seemed to get a real kick out of how we dressed. In fact, I broke out a second bandana, which they took turns wearing. A few of the kids did dead-on impersonations of Ron and me. We even went to a school dance. That was a lot of fun, but nothing beat our interaction with the students in the classroom. They were hungry for knowledge, and we loved spending time with them.

Actually, our experience in clinical settings proved a great help in getting mummy guardians to trust us. I think it provided us with a level of compassion that was valuable in our work with human remains. Occasionally, we have to ask patients to do things that are uncomfortable or unpleasant. You really have to win their confidence before they will let you do that. We often had to convince the mummies' caretakers—in many cases, their relatives—that we would be treating the mummies with the same respect we give to our patients. I think this came through even when there was a language barrier.

The fact that *Mummy Road Show* aired on the National Geographic Channels also kept us focused on the cultural component of what we were doing. Everyone was aware that the show had to fit the National Geographic theme. This was fantastic for us, because we would go to these places and experience different cultures not as tourists, but from the perspective of the people with whom we were interacting. It also underscored for us the ways in which the current-day culture related to the mummies we were investigating. The mummy is part of the thread that connects who they *are* to who they *were*, and it's really interesting to be a part of that.

The whole *Mummy Road Show* experience also made us realize how Westerners are often truly ignorant about a lot of stuff that's happening in the rest of the world. Think about the problems this attitude creates for Americans around the globe. You can't just show up in a different culture and impose your standards on people who have been doing things a certain way for hundreds—or maybe even thousands—of years. They won't accept that. This has a major impact on the study of mummies. There are great mummies all over the place, but we are unable to study them for cultural, political, or religious reasons. Embracing different cultures is an important first step in breaking down these barriers.

Steamy Stuff

Ron and I were always eager to experience the local culture, but we should have known something was up when Larry Engel insisted that we go with him to the Turkish baths while we were filming "Turkish Tomb Mummies." The trip started with an escort of another kind; the local tourism department was very excited to welcome us, and had arranged for a police escort from the airport to our hotel. We were then greeted in an official ceremony. Ron and I were exhausted from the flight, but there was plenty of coffee to keep us awake.

When it came time for the Turkish baths, we almost wish we had slept in. We'd experienced saunas, of course, but neither of us had been to an authentic Turkish bath. So we said, "Sure, let's do it." It was an amazing place. It looked like it had been there for centuries. We enjoyed a nice steam, not realizing that the most important part of a Turkish bath is the deep massage that comes later.

You are brought over to this stone table that's glistening with sweat and steam, and before you get up on it they hose it off with water. Then they start working on you. Man, was that painful. It was as if they didn't think they were doing their job if you weren't screaming. No matter where they were working, they just kept pushing until they reached bone. At least Ron and I knew which muscles they were shredding. Did we feel better afterwards? I'm not sure. I do know, however, that the nicest part of the massage was the Fanta orange drink handed to each of us at the end.

The work of the *huaqueros* in Ilo, Peru. This is where we filmed the "Mummy Rescue" episode.

Another interesting aspect of getting to know the people in the areas we were working is that we acquired a much deeper understanding of some of the issues we had previously seen as being kind of black and white. Looting, for instance, is an appalling concept to a scientist—particularly when you work with mummies. In Peru, everything we know about those cultures comes from tombs. Unlike the Egyptians, the Chiribaya of southern Peru (a culture that predated the Inca) did not build their tombs with stone. There is little evidence left of their villages and no written records, so what we find in the bundles, such as textiles and artifacts, provides crucial information regarding the culture.

Yet looting goes on all the time. People take things and sell them to feed their families, just to survive. Almost anywhere tourists might be found, you can bet that someone will approach you with artifacts for sale. The looters in Peru are called *huaqueros*, and there are lots of them. Some anthropologists and archaeologists actually hire these folks to help them with their digs, partly for their expertise and partly in an attempt to maybe cut down on looting. Does this mean they are dealing with the devil? It depends on how you look at it. These scientists figure that if grave robbers are paid, they'll probably protect the site for them.

Since I started going to Peru in 1997, I have begun to understand the bigger picture of looting. First off, there is high unemployment, so everyone is trying to make a buck. Second, the indigenous people watch as archaeologists and anthropologists who appear to be much better off than they are take these things away with them—presuming that they are doing this to make money. And third, this is a culture that has watched people dig up and destroy their ancestors' graves and mummies since the Spanish arrived. For five hundred years, they have been told that their non-Catholic ancestors were basically worthless. The desecration of their tombs has become part of the culture. So what could possibly be wrong with taking a few things to put food on the table?

It is heartbreaking to go to a place like El Brujo, an archaeological site on the north coast of Peru where the ground is covered with the skulls and bones of looted mummies. Since the skeletons are worth nothing monetarily, they are simply discarded by looters. You also see bits of textile, because unless an ancient textile is perfect, it won't sell. If you go to the market in Lima, you will find beautiful pieces of textile under glass for next to nothing. The looters are looking for the artifacts in the mummy bundles. That's where the so-called big money is. The real tragedy, of course, is that the artifact they get $20 for might sell for hundreds of dollars, or even thousands, once it reaches the collector's market.

The hope for the future is that the current generation of kids in Peru is being educated by people like Sonia Gúillen, a bioanthropologist who we worked with on "Mummy Rescue" and "Holey Mummy." She hopes to get them to understand and appreciate their heritage, with the idea that they will not become huaqueros themselves.

Of all the episodes we did during *Mummy Road Show*, the one that really brought all the different themes in this chapter together was probably "Cave Mummies of the Philippines." We had worked in tombs in isolated places, but this location in the Kabayan Mountains was really

Discarded bones at El Brujo.

out there. What a remarkable experience! In fact, getting there was part of the challenge. Indeed, it was the rainy season, and landslides made the road to the Kabayan village virtually impassable. But we persevered, and were we ever glad we did. Our guide, Orlando Abinion of the National Museum in Manila, was fantastic.

The community lined up a whole procession of people whose responsibility it was to make sure we treated their ancestors appropriately. We had the approval of the town's mayor, but the town's elder, Baban Barong, was still the *de facto* boss, and we needed his blessing. Three pigs were sacrificed, and the village shaman consulted the liver, and then gave us permission to work on the mummies. When we first saw one of the livers, we thought, *That's gonna be a No.* There were notable lesions on the surface. Luckily, the spiritual leaders quickly flipped it over and the other side was clean. Anyway, the whole village ate well that night.

The deal with the villagers was that we couldn't take the mummies outside of the caves, which were carved into huge boulders. We love a challenge, especially a technical one, so that was fine. Fortunately, we had two Filipino army guys to carry the generator and the X-ray machine up the mountain. We were eventually told that they were there to protect us—there had been some guerrilla action in the region. A year later, a bus was blown up on the same road we had traveled. It was probably easier for us to work not knowing the potential danger.

There were other reminders of just how far from civilization we were. We were told that we were among the few outsiders besides the Japanese army to come through there in a century. What was most striking about how people lived in this remote area was the reverence they held for their ancestors. They really, really cared about these mummies, and unlike the people we had met in Peru, there had never been a disconnect. This was fascinating to us, because the villagers were predominantly Catholic. But being so far off the beaten path, they had been able to blend Catholicism with their own belief system in a way that seemed to work. Otherwise, who knows what would have happened to these incredible mummies.

Before we x-rayed each mummy, the villagers accompanying us would perform a ritual, and Baban would say a prayer. We sensed it was a very spiritual process and asked what was being said before we went to work on a child mummy. Our suspicions were confirmed. Baban was communicating with the ancestors, telling them that we were going to x-ray the child's head. He also reassured the child that we would be careful.

There was a tremendous amount of spirituality surrounding these people and their relationship with the mummies. I found it very touching. They wanted to convey to the mummies what we were about to do. When the wind blew through the valley, it whistled through the caves; the villagers believed this was the voice of the mummies up in the mountains—their ancestors—speaking to them.

One thing that really impressed us was how interested the young people of the village were in what we were doing. They were just like the elders in that regard. They watched us, took notes, and always seemed to have a drink for us when we got thirsty.

Because the work in the caves took two days, it was decided that the entire group—villagers and film crew included—would stay on the mountain overnight. Instead of cramming into the small shelter on top of the mountain, Larry Engel, Ron, and I chose to spend the night outside by the caves. We had only sleeping bags, no tents, and it rained a good portion

The remarkable Baban Barong.

of the night. By the morning we were drenched. But that's when we experienced the best part of the trip: below us in the valley, a rainbow appeared. There's something very special about looking *down* on a rainbow, especially when you are immersed in a culture that is always looking for omens. This was a positive sign for the locals, and even though we are scientists, we like to think that we had something to do with it.

Where the Dead Things Are

During the early 1970s when I was doing research in comparative osteology at Yale, we were always on the lookout for bones of different animals. Over a ten-year period, the group of scientists I worked with had assembled a number of diverse sources for dead animals. Instead of having to resort to euthanasia, we picked up carcasses from farmers, meat packers, animal dealers, and even other research facilities. Research teams would collect animal specimens from all over the world and ship them back to facilities in the United States. We were so successful in acquiring material that the Orthopedic Department's walk-in freezer was referred to as "Noah's Ark."

If scientists didn't properly preserve the specimens they collected, the resulting decomposition would create what most people would consider a rather unpleasant mess, which was of little value. However, since my interest was in the skeleton, I was quite happy to accept this type of material. This was the case with two 55-gallon plastic barrels filled with seals that had been shipped to a facility several years earlier. As it so happened, at the same time I picked up the two barrels, an animal dealer had two frozen emu carcasses available. The skeleton of the ostrich-like birds would make a fine addition to the collection. How could I say no? Finally, to add to this mixed grill was a frozen immature dolphin from another research facility. The marine mammal was small enough to allow me to get great X-rays, and fresh enough to permit a necropsy.

I decided to bring everything to the apartment I had just leased near the University of Connecticut. The barrels were easy; I put them in the carport. But how to thaw the remaining material? The dolphin fit into the bathtub, and once I filled the tub with water, the thawing process proceeded at a controlled rate. Because the emus were large and feathered, I decided to thaw them outside. However, I didn't want to attract dogs, cats, or other creatures that resided in the nearby woods. The roof seemed the logical choice, so I tied them to the bathroom stand-pipe. It was autumn, and since they were on the north side of the roof, it would take a day or two before they thawed and I could start the dissection.

Oh, did I mention that I lived with my girlfriend, and she had no idea what I was bringing home? I decided to start thawing the seals before she came home. After prying the lid off a barrel, I got my first look inside. The best way to describe it was a kind of chocolate-colored, soupy, layered mess. The odor, on the other hand, was beyond description. Once inside the barrel, it was possible to gently ease each amoeba-like mass that represented a seal onto a sheet of plastic. After removing the animal from the container, it was a simple procedure to extract all of the bones and put them aside to dry. I placed the thick liquid and chunky bits that remained in a bucket. I emptied the bucket in a hole previously dug behind the apartment. After about three bucketfuls, I covered the opening with dirt and dug another hole.

Before I had finished the first barrel, my girlfriend arrived home and got her first look at my "zoo." She wasn't pleased, to say the least. It didn't get any better when she went in to take a bath and found the surprise in the tub. We stayed in that apartment until spring. By the time we left, many of the neighbors were curious about why their leashed dogs would drag them to this certain area behind my apartment. Huge masses of flies also seemed to congregate in very specific patches. I never said a word.

Possibly the grossest thing that ever happened to me took place during this period. Dr. John Ogden, chairman of Yale's Orthopedics Department, Alden Mead, a research associate in ophthalmology, and I had established the Yale Marine Mammal Stranding Center. We had a permit to work on beached animals such as whales, dolphins, and seals that died along the Connecticut coast. As part of the agreement, we also examined dead marine turtles. On a trip back from the Smithsonian, where I had picked up some specimens, I made a detour to Atlantic City to see Bob Schoelkopf. Bob, who was director of the Marine Mammal Stranding Center in Brigantine, New Jersey, had called me about a dead leatherback turtle he had. Since we had a turtle expert, Dr. Anders Rhodin, at Yale, and strandings of this type of turtle were not common, it seemed only logical to bring the leatherback back with me.

Leatherbacks get to be quite large, sometimes over a halfton, and Bob, like most of us in this business, was an expert at adapting equipment to meet specific needs. He had converted an old fire truck to serve as a hoisting mechanism to get the stranded sea animals up off the beach. It was well after dark when I pulled into the Stranding Center. I took one look at the enormous turtle and asked Bob how long he thought it had been dead. "Oh, not that long," he said. "It's not that bad."

Since the turtle was already on a sling, Bob hoisted it up with the modified truck. I drove up beside the animal and opened the two side doors of my van. We got the 1,200-pound behemoth swaying and maneuvered it into the van. With the help of a couple of two-by-fours, we were able to close the doors.

I bid Bob farewell and began the five-hour drive back to Connecticut. It was probably around nine in the evening by the time

I got to the Garden State Parkway in New Jersey. This was during the 1978 fuel shortage, when you could only buy gas every other day, depending on whether you had an odd or even license plate. I had the wrong plate for that day, and there was no way I was going to make it back home on the gas I had left. No problem. It was getting late, the needle was on empty, and I was exhausted, so I decided to pull into a rest stop, take a short nap, and wait until after midnight to refuel.

At around two in the morning I woke up with a strange feeling. I was parked near a light, so the interior of the van was illuminated. When I looked down, everything seemed to be moving. Maggots—hundreds of thousands of them—were just pouring out of this turtle, and they were covering me. I freaked out. I leaped out of the van and tore all my clothes off. I was jumping up and down in the parking lot, trying to get the maggots off. Finally I regained my composure, and cleaned out the van as best I could by scooping the maggots with my hands. I got dressed again, bought some gas, and got back on the road. Waves of maggots were still coming out of this turtle, so at every rest area between New Jersey and Connecticut, I had to scoop them out of the van. That was something else!

I still have a souvenir from those bygone days: a horse skull displayed proudly on top of a bookcase in my office. The source was a meatpacker in Hartford (it was a full horse head back then). At the time I was working in the radiology research lab at Yale, doing comparative neuroanatomy. I had a small lab that had roof access, and the roof seemed like the perfect place for temporary storage of the horse head, which I placed in a large plastic container. I must have left it there a little too long, because a stream of maggots crawled into a vent that led directly to the offices and labs on the floor below. Someone was apparently sitting at a desk when maggots began dropping from the overhead vent. I'm a quick learner, and that never happened again. From that point on, all material that was going to be skeletonized went to the Peabody Museum, where the dermestid beetles removed all the flesh. But that's another story.

Chapter Four

Mummy Mementos— Ron Beckett

A nyone who watched *Mummy Road Show* knows that Jerry and I get really excited when we find some sort of artifact associated with a mummy. These "mementos" accompanied people on their journeys from the world of the living to the world of the dead. Sometimes they have been thoughtfully or lovingly selected for burial, sometimes not. Either way, they can tell us a lot about the individuals we are studying.

Sometimes a mummy memento helps us turn a corner. The nail up the nose of the Carnival

A mummy memento from Cajamarquilla, Peru. This beautifully woven textile, featured in "Mystery in a Bundle," was meant to accompany the deceased from this life to the next.

Mummy suggested he was a blockhead who had passed away surrounded by other sideshow people who respected and understood him. At other times, a mummy memento can turn a case on its ear. Like the buttons on the Soap Lady in Philadelphia. She was assumed to have been a victim of a cholera epidemic in the late 1700s, but the buttons on her clothing proved otherwise. We had an expert at the Smithsonian narrow the buttons down to about 1828. He could even name the special machine that had produced the button.

When it comes to mementos, one question we hear often is: Have we ever taken something we've found with a mummy as a souvenir? I understand where this comes from, but we have never even been tempted. There's no turning back once you go down that road. Don't get me wrong; it's human nature to want a keepsake from an unforgettable experience, something that jogs the memory when you touch it or see it, something that reignites the fire again, reminding us of where we've been.

I have to say, we know people with quite a few artifacts in their homes, and it is nice to see them, without a doubt. If you come to our homes, however, you won't find anything like that. It's not that we don't enjoy mementos from our travels. Rocks and shells are among our favorite keepsakes, as are assorted *tchotchkes*—like the little souvenir mummies made by the nuns at Leymebamba. But never bones or artifacts or textiles. I happen to like feathers, so when we go to different places, that is what I look for. Sometimes I will forget and Jerry will bring one home for me. I always appreciate his thoughtfulness. Feathers remind me of the spirit of a place. I also like sand, for the same reason. When Jerry wanders a site and sees an interesting-looking rock or stone, that's what he is most likely to take.

Have we been offered artifacts after filming an episode? Not that I recall. We work with reputable archaeologists and anthropologists, and this isn't something they're likely to do. They have their reputations to think of, too. On the other hand, we have been offered stuff to buy from the locals—everything from legitimate artifacts to folk art to traditional clothing. For many of these people, especially in South American countries, it's an economic thing, and you truly come to understand that. Still, generally we say no, because you almost feel like you are perpetuating some sort of exploitation.

But this isn't always the case. An example is Jenny Figari de Ruiz, who is the director of the Yachay Wasi Institute in Lima, Peru. She has this great thing going, where the local women do traditional weavings and receive a fair wage. In this way, Jenny is trying to preserve the past by bringing it into the present. Meanwhile, not only do these women generate an income, but they also earn a great deal of self-respect.

Would I want to own a mummy? No. Would Jerry? I doubt it. Besides, I've been to his house—he'd probably lose it.

We made one of our coolest finds during the filming of "Tales from an Italian Crypt." This episode was great because we got to work with one of our favorite people, Gino Fornaciari. Gino is friendly, cooperative, and an excellent scientist—and did I mention that he's brilliant? Gino is a pathologist, so he's used to cutting and slicing and dicing. His impressions of our X-ray and endoscopy work were very positive; he was amazed by how much information we were able to gather with our non-destructive techniques.

In "Tales from an Italian Crypt," Jerry's X-rays revealed what appeared to be a coin inside a mummy. I went in with the endoscope, found the pouch it was in, and removed it. It turned out to be a medallion, and its significance was explained by a note with a piece of clothing wrapped inside of it, both also located in the pouch. The image on the medallion was Saint Philomena, a thirteen-year-old who declared herself a bride of Christ in the third century and was then beheaded when she refused to marry the Roman emperor, Diocletian. The medallion and clothing swatch were intended to help this individual's safe passage into the afterlife.

Obviously, these artifacts were of tremendous interest to us, particularly because this mummy was proving somewhat difficult to date. By looking at the elegant garb he was dressed in, we were able to place him in an era, but that was as close as we were going to come with the evidence that was available when we arrived. It was the medallion that cracked the case. Philomena was declared a saint in 1802, and we were consequently able to determine that the medallion had been made some time in the next eight years or so. This told us that this gentleman could not have died before this time. In the end, this episode demonstrated the value of finding an artifact like that.

One other footnote from this episode was a wonderful line delivered by Gino, maybe the best one we heard during the entire run of *Mummy Road Show*.

Gino was commenting on the medallion we had discovered in this little pouch, and with his beautiful Italian accent, he called it a "passport to paradise." What insight and perspective! Jerry and I looked at each other and realized instantly why we loved working with Gino. No one has more style or grace.

On the "Pirate Island" episode, on Isla San Lorenzo, Peru, it was also a religious medallion that helped nail down the time frame for a particular group of mummies. In this show, we were hoping to find the remains of legendary buccaneers who were part of a Dutch force intent on invading South America and eliminating the Spaniards from the New World in the 1600s. This time around, however, it was not a case of tweaking that number by a decade one way or another. The medallion discovered with one of the mummies was proof that this group was from an entirely different era.

The fact that they were located on San Lorenzo, an island off the coast of South America, provided an initial clue that they could have been pirates. Also supporting this belief was the condition of the remains, which showed signs that this man had been extremely skinny.

The first three mummies we encountered were adult males. The fact that they were located on San Lorenzo, an island off the coast of South America, provided an initial clue that they could have been pirates. Also supporting this belief was the condition of the remains, which showed signs that this man had been extremely skinny. This suggested they could have suffered from dysentery, a very common disease among sailors of this period. San Lorenzo would have been a place for the Dutch pirates invading Peru to bury their dead.

But the medallion we found with one of the mummies shed new light on our subjects. The first set in this series of medallions was struck in 1832 in France, so there was no way the buccaneers of legend could have had one. In this instance, a memento was crucial to uncovering the true story behind a specific group of mummies.

Later during this episode, we discovered the remains of women and children, too. It turns out they were probably European settlers who died of cholera—a disease that also causes drastic weight loss—and were buried on San Lorenzo. What was interesting is that San Lorenzo appeared to have been a kind of makeshift Ellis Island for Peru—and in a way, that is a much more interesting story than pirates. We just happened upon this graveyard that was probably from the 1840s. Europeans coming by ship would have been quarantined on the island for a while, and the ones who died would have been buried there. We also found some indigenous burial sites while we were working on San Lorenzo, indicating that there is a lot more work to be done there.

Half the fun of what we do is meeting other people just as intrigued by mummies as we are. And a mummy memento almost guarantees we are going to make some interesting new connection. Over the years we have come to know a group of people who we can reach out to, or who can point us to someone with the unusual expertise we need. In "Medical Mummies," we worked with Ronn Wade (we both love the double n's) at the University of Maryland. He found a tattoo on one of the mummified skin specimens in their collection.

We were able to consult, via the Internet, with two experts— one on tattooing and its history, and the other on the symbolism of the time period we believed this arm came from.

The tattoo spelled out the name of Pope Pius IV, but the other symbols were a mystery. The experts were able to tell us that these symbols also related to Pius IV. It was a controversial time in Christianity, with churches vying for power in

The tattooed arm of a mummy at the University of Maryland School of Medicine.

England; with the experts' help, we surmised that this person (a prisoner, we believed) had this tattoo as a way of showing his dedication to Catholicism. We were also able to determine that the tattoo was probably done by a professional tattoo artist of that time, which surprised us. We thought it looked more like graffiti. Of course, we were judging by twentieth- or twenty-first-century standards.

Fascination with Plastination

The "Medical Mummies" episode was a great introduction for us to Ronn Wade, the keeper of the cadavers at the University of Maryland School of Medicine. Ronn has tremendous respect for the science and study of human anatomy in all its variations. His collection of medical mummies there is fantastic; it ranges from salt- and sugar-cured specimens all the way to the most modern techniques of plastination.

Plastination is one of the most up-to-date ways of preserving the human body for medical study. The beauty of it is that you can take any part of the body and plastinate it—a heart, a larynx, or an entire extremity. Ronn showed us the process, and seeing it was an eye-opening experience.

Basically, the human body is 70 percent water. In plastination, the water in each cell is replaced with acetone. The acetone is then replaced with silicone, which serves to preserve the target organ indefinitely. The good thing about plastination is that the shape and pretty much the color of the body part is retained. In teaching anatomy, color is important, texture is important, shape is important. With Formalin and other types of preservatives, you lose these characteristics. Plastination enables students to study a complicated organ like a larynx in ways they were never able to before.

The whole idea is to look at an individual part, perform a dissection, and then plastinate it. Plastinating an entire body for the purpose of scientific study would be wonderful because the spatial relationships between and among structures could readily be explored.

A controversial German anatomist, Dr. Gunther von Hagens, has plastinated bodies and then exhibited them on tour. People feel that he crossed the line between anatomy and art. Bodies donated for science were never intended to be displayed this way.

We also encountered tattoos when we shot "Cave Mummies of the Philippines." This episode was a blast for a number of reasons. As Jerry mentioned previously, we had to win the hearts of the people who were going to take us to the mummies, and we also had to win the favor of the gods. Our guide, Baban, had helped the Americans during World War II when the Japanese occupied the island. He must have been eighty years old when we met him, but he went up that mountain where the mummies were buried like Spider-Man.

There was one particular mummy that had incredible tattooing. We were told it had been done with a mixture of ash and tomato juice. They would use a thorn to hammer a pattern and then rub this stuff in. The tattoos weren't just little geometric patterns; they were incredibly complex, with both animal and human figures. It must have been a very long, involved process. We were also told that the more tattoos you had, the higher

The tattoos on this individual's legs suggest he was a person of high status.

your status. One particular individual had tattoos on his legs, arms, back—it was amazing. He must have been very important. In this mummy's case, the tattoos could not tell us much else beyond his social status. There is a lot of work to be done on this culture, and hopefully someone will be able to someday link the tattoos to specific times, beliefs, or events.

We were told the mummies were from the ancient Ibaloi culture of the Philippines. There are still some Ibaloi bloodlines in the village of Kabayan, near the mountain, and it is certainly possible that this story is true. These areas are isolated in the mountains, so Spanish blood probably did not mix with theirs until the 1800s. We could not be sure how old the mummies were. They did not let us take samples for carbon dating, although these samples probably would have been contaminated. Apparently, these mummies were occasionally removed from their caves, which can alter the test results.

One of the best mummy mementos we ever found was a crotch rivet. The episode was "Mummy in Vegas," and we were investigating remains that had been unearthed by a private builder in Carlin, Nevada. Had it been a commercial developer on a budget and a timetable, it might have turned out differently, but this guy really felt like he had been given the opportunity to make a real contribution to history, so he delayed construction while the site was excavated. It turned out to be a cemetery for Chinese immigrant workers. Anyway, while we were x-raying one of the mummies, we saw a rivet that was most likely from a pair of Levi's. Jerry and I wear jeans, we work in jeans—heck, we even *sleep* in jeans—so it did not take us long to figure out what it was we were seeing.

Although Levi Strauss has been making jeans for more than 100 years, we did some research and dis-

Though a bit corroded, this special find from "Mummy in Vegas" was one of our most riveting discoveries.

covered that the company only made a crotch rivet for a certain amount of time. When cowboys were sitting around their campfire at night eating their beans, the crotch rivet would heat up, with the predictable result. This nugget about Levi Strauss helped us narrow down the time frame on the rivet, which put this fellow in this part of Nevada right around the time when the railroad was coming through.

Another interesting thing about this mummy was that he was clearly Chinese. We knew this from his hairstyle (a long braid) and from some Oriental-style robes he wore. Yet he was also wearing his jeans; here was someone clearly caught between two cultures. Once we had this information, we were able to look at the other evidence and piece together what was probably a pretty hard life as an immigrant worker.

It is funny how something so small can be so exciting. No one gets more excited about these types of finds than Jerry. As soon as he recognizes the shape of something in an X-ray, you can see it in his face. He's

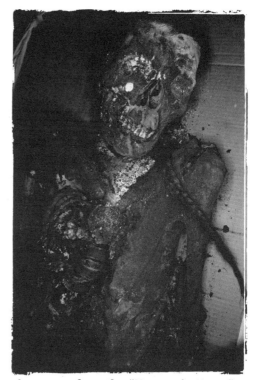

A mummy from the "Mummy in Vegas" episode of The Mummy Road Show.

like a kid at Christmas shaking one of his wrapped presents. *Mummy Road Show* did a great job conveying the enthusiasm we felt in these situations, but I always wondered whether any viewers got as excited as we did when we found something like a rivet. Part of that excitement was in knowing that this artifact would open the door to anthropologists, archaeologists, and other experts, possibly allowing them to reconstruct a lot of the details of this person's life.

By the same token, one of the frustrations we had while doing the show was that we had a very small window of time to work with the mummies. Your instinct as a scientist is to follow the trail you uncover, but that was not always possible on *Mummy Road Show*. We had thirteen weeks to shoot an entire season, so we were always on the move. The first season in particular was very hard for us in this respect. Stopping what we were doing and leaving was not an easy thing to do. We eventually learned that we just had to back off. Now Jerry and I are taking another look at many of the investigations we started, and saying, "Gee, it'd be nice to go back and work with some of these mummies, and some of these scientists we met, and really take our time and do more thorough work."

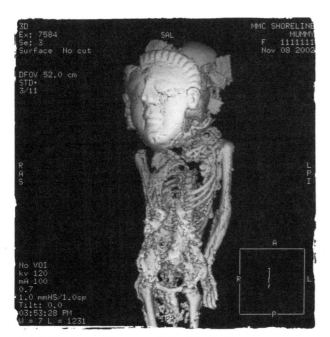

A 3-D CT from "Mystery of the Masks."

Hopefully, the people who watched the series understood that what we were doing was a survey, an overview. Ideally, we would turn over everything to people who had the time and resources to do the proper study, like the folks at the University of Las Vegas. We introduced them to our viewers in "Mummy in Vegas." At least one dissertation was already under way by the time we got there, but others were launched with the help of the work we did, and the mummy we examined is now a topic of study for graduate students. That's a nice progression. This is particularly important, because some of the people buried in the Carlin cemetery may have been among those who built the railroad—which transformed this part of the west.

Sometimes a mummy memento can lead to a reunion. This occurred in "Mystery of the Masks," which was based on work we had started before the series began filming. We were at the Peabody Museum of Natural History at Yale University looking at bones to determine the age and sex of a mummy, and hoping to shed some light on any diseases it might have had. While we were working, we also became intrigued by the mummy's mask. It had some Christian motifs, as well as Egyptian beliefs woven into the iconography.

While we hadn't initiated this study, this didn't dampen our enthusiasm. An expert had determined the region of Egypt the mummy came from, the time frame, and—this was really neat—the fact that it may have been part of a group of mummies looted from the same location and subsequently split up. One mummy was at the Metropolitan Museum of Art in New York, another was at the British Museum in London, and possibly one in Edinburgh, too. The placement of certain images on the masks suggested that they were done around the same time, in the same region, perhaps for the same burial, and by the same group of artisans.

As usual, we also needed to do some X-rays. We had a long-standing relationship with the Peabody Museum, so it was easy to accomplish our task there. The folks at the Metropolitan allowed us to do the X-rays, but they were extremely protective and would not permit us to make contact with their mummy. Fortunately, Jerry is one of the few guys who could get the X-rays we needed while working under these restrictions. The fellow with whom we worked at the British Museum was wonderful, but the museum itself also had a lot of rules we needed to follow. This did not surprise us. Museums in general tend to be somewhat territorial, and are very protective of their mummies. They are not only concerned about any possible damage to the mummies; they are also concerned about what you plan to do after you leave. They don't want to see a picture of their mummy in a beer commercial. Understandably, they want total control.

As it turned out, the mask at the Yale Peabody Museum was a really close match to those in the other two locations. It appeared that all of the masks were from a narrow time period and perhaps from the same place. In fact,

A painted cartonage from the Metropolitan Museum of Art in "Mystery of the Masks."

The Yale mummy gets a CT scan.

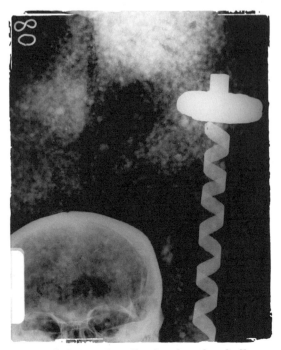

X-ray of an Inca bundle showing the skull of a mummy and a mace.

one was so close to another that we wondered whether it was produced by the exact same artist!

The mementos we found in Peru while filming "Mystery in a Bundle" were really spectacular. They weren't unexpected in and of themselves; what was unexpected was the quantity and quality. In the case of this one individual, whom we called the Lieutenant, the shaving tools and textiles suggested he had a fairly high station in Incan society. This was important, because the mummification was fairly poor. The endoscope was able to show a few things, but in this case, the X-ray was a wonderful tool for revealing artifacts. The presentation of this mummy was particularly interesting, because archaeologist Alfredo Navarro was there. He had a good handle on the history of the site and the different cultures that had occupied it.

We investigated another large bundle in Puruchuco for "House of Bundles," and it contained a scepter, among other items. This was a late Incan site, too. The scepter was an important find because it indicated that the mummy it was buried with likely enjoyed a high status among this huge group of people.

The textiles leave tantalizing clues about the person they are wrapped around. A typical bundle looks big enough to hold two or three mummies. Another thought that crosses your mind is that the bundle must be full of artifacts. But there is usually only one mummy in a bundle, relatively few items, and a lot of wrapping. I remember how frustrated we got during "Mystery in a Bundle." Jerry's equipment was stuck in Miami, so he had to use this old dental unit for X-rays. (In fact, every time he fired up this machine, the village nearby would experience a brownout.) Seeing inside this bundle became really difficult. Fortunately, Jerry was able to get the shots we needed.

In many bundles found in Peru, it is not unusual to find textiles with 200 threads per inch. With cloth like this, each bolt must have taken hundreds of man-hours to make, which is noteworthy, because time was an important currency in ancient cultures. Were these mummies people of importance? Could they be ordinary folk? Unfortunately, we may never know. The early missionaries destroyed so many mummies in Peru, and the chroniclers at this time were Spaniards, who were not exactly embracing this culture they had conquered. Mummies were being burned, not catalogued, so a ton of crucial information is probably gone forever.

Even in the best of circumstances, Peru can be a tricky place to work, especially when it comes to dating stuff. This is why you really get jazzed when you find a memento of some kind. In fact, without some type of artifact, there may be no way to put a mummy in temporal context.

Parts of Peru are so arid that anything that goes into the ground becomes a mummy within weeks. How dry is dry? We once worked with a veterinary anthropologist—this wasn't on the series—who was looking at parasites on dog mummies. She brought this big dog back to the laboratory in Ilo, and had been working on him for a couple of weeks when she made an amazing discovery. It was Sonia Gúillen's dog, Fujimora! Sonia was the director of the research center there; Fujimora had died a year earlier and was buried in the sand. To Sonia, Fujimora was like a family member—a dynamic that we would gain a deep appreciation for throughout our travels on *Mummy Road Show*.

The Eyes Have It

One of the weirdest mementos we encountered belonged to the blonde nun we investigated in Turkey. She left us her eyeballs, perfectly preserved. It was bizarre, because you actually

It is difficult to describe the experience of looking a mummy in the eye, as we did in "The Blonde Nun."

could see the color of her iris. That is significant, because she would have had to have dried out very rapidly for the eyes to be preserved like that. If you don't have rapid desiccation—we are talking about a matter of days—the eyes just collapse.

It is a different experience when you can see the eyes in a mummy. As a rule, mummies do not look like people do. They are dehydrated, they are shriveled up, and their mouths appear wide open. They are recognizable as humans but not as people. But everyone can identify with eyes and eye color. There is no way to describe the experience of finding yourself looking right into the eyes of this person who died a thousand years ago.

She was a looted mummy that had never been studied before. We were fortunate during *Mummy Road Show* to get first crack at a lot of mummies.

The other interesting thing about this woman was her blonde hair. We sent the hair out for isotopic analysis, and it showed that her diet was primarily seafood. Well, there was no seafood in this region of Turkey—it was the middle of the desert. It would have taken a couple of days for anything to reach this location, and seafood would not have been in the best condition after that type of journey.

She probably came from someplace else and died, or died someplace else and was transported there. She may have come from a port city, where her seafood diet would have made sense. Whatever the story, the eyes of this mummy reinforced an important fact: All mummies were once living, breathing individuals who had hopes and dreams and went through trials and tribulations, just like the rest of us. It may be a cliché, but it's true; the eyes are indeed the windows to the soul, especially when they belong to a mummified blonde nun.

Chapter Five
Like Family—
Jerry Conlogue

Y̲ou know how you always hear about scientists not getting emotionally involved with their work? Well, to a certain extent this is true. However, *Mummy Road Show* created a uniquely intense situation for Ron and me. The goal of each episode was to find out as much as possible about the mummy we were studying. Because of the tight filming schedule, we had to throw ourselves into our work. For as many as twelve hours a day, we were totally immersed in our research. Under these circumstances it became impossible to remain detached for long. The more you discover about how people

The late Captain Harvey Boswell, who owned Marie O'Day.

lived and died, the more you can identify with them. In a bizarre way—especially for mummy caretakers—the preserved remains become like a member of your extended family. This is definitely true for the people who are around them all the time.

Ron and I often discuss how this is one of the most fascinating aspects of what we do. The people caring for mummies are really interesting. You have people who treat mummies like relatives; you have caretakers in museums who want to protect and learn from them; you have researchers in the field who want to know more about their mummies in a collaborative way; and there are those for whom a mummy is a cash business. You have people who want to be around mummies as a career choice, and some who become accidental caretakers.

Some people are extremely protective of their mummies; some are willing to pay to have scientists work on their mummies; others are looking for a little payment for the privilege. What they have in common is that they are all passionate about their mummies. Even Hazel Farris's owner, whose only wish was to have her cremated, wanted to get her story straight before she put Hazel in the ground.

Parental Guidance

The toughest mummies to work on are kids. Ron and I both have children, so we find ourselves empathizing with the parents. We imagine what they must have gone through in losing an infant or child. We can't imagine anything more painful. This point is driven home, often in a heartbreaking manner, by what we find buried with kids. On Pirate Island, off the coast of Peru, we found the remains of children surrounded by little flowers and mementos, and that was incredibly touching. At Guanajuato, we also saw a lot of child mummies, all dressed in fine clothing. Prior to the advent of antibiotics, infants and children were much more susceptible to infectious diseases, which led to a high mortality rate. The reason for dressing up the children may have been that grief-stricken parents believed that this would ensure admittance into heaven. This was a custom of the period, as was funeral photography.

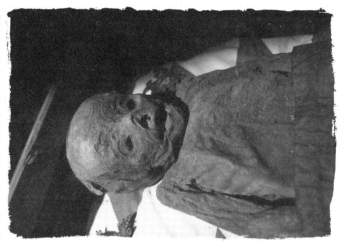

Death is viewed differently in different cultures, but infant mortality and the death of children always seems to carry more meaning. It was true

Princess Anna, a real heartbreaker.

a thousand years ago, and we see it in the eyes of people visiting child mummies today.

In "The Princess Baby," the caretaker of Princess Anna was genuinely concerned about her in a parental way, although I suspect he would have been just as caring with an adult mummy. Anna was kind of like the child of the entire community of Kastl, Germany; they had created a whole festival around her father, King Ludwig, and the fact that Anna had ended up in their monastery only strengthened the connection.

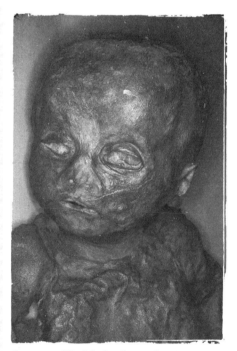

Anna was interesting emotionally for Ron and me, and I know that feeling came across on camera. We saw the telltale signs on her X-rays that indicated she had suffered repeated bouts of illness when she was young—something I could really identify with. Our hearts just went out to this little girl. Not only was she young when she died, but

A mummified baby from the Amasya Museum in Central Turkey. Ron and I worked on her in "Turkish Tomb Mummies."

she had also endured a lot of suffering while she was living. In these situations, sometimes you have to step away for a second.

This show was one of those times when we said something from the heart that Mary Olive Smith, the episode's director, really wanted to get just right on film. She would often ask us to repeat something profound if the lighting or sound had been less than ideal. But of course, we usually couldn't remember exactly what we had just said, and the moment would be lost. We would try our best, but we're not actors; inevitably, it was never quite as good the second time around.

Sometimes mummies are family heirlooms; like Hazel Farris, they are passed down through the generations. In "An Egyptian Souvenir," the mummy was handed down from generation to generation within the Sulman family of Chatham, in Ontario, Canada. This was our very first episode, the beginning of a process that was very new and very different, and it was a great one to start with because you could see this family really cared about this mummy. The old man was trying to pass that love down to his grandchildren. He had taught himself how to read some of the hieroglyphics on the mummy's mask, and he talked about how he used to play with the mummy as a child. We really felt close to that mummy, as well as to the caretaking family.

Chatham is a small community, and over the years the Sulman family has done a lot to bring the world there. It was like turning back the clock to a time when people wealthy enough to travel abroad returned with strange and wonderful souvenirs from their adventures. The Sulman family thought enough about their community to donate artifacts brought back from Egypt decades ago to the Chatham-Kent Museum. It was not on the same scale as an Andrew Carnegie, but the same thought went into the gesture. And in doing so, the Sulmans created a wonderful legacy for their family name.

This mummy had been x-rayed before, but in the three or four years since we did the show, it has undergone a CT and a number of other non-destructive tests, and has spawned interest within its own small scientific industry. I feel really good that we kicked that off. There is an exhibit now that incorporates the work we did, similar to the Soap Lady exhibit at the Mütter Museum. This was the first of many instances where *Mummy Road Show* ignited an interest that others have carried forward. That is a wonderful legacy for us.

One might categorize these mummies as "curiosities," but the people caring for them did not feel this way.

Where sideshow mummies were concerned, I think one of the important things the *Road Show* pointed out was that there was never anything disrespectful about the people who owned them. One might categorize these mummies as "curiosities," but the people caring for them did not feel this way. They really did care a lot. Marie O'Day—whose owner, Captain Harvey Boswell, passed away shortly before we filmed the "Wonder Mummy" episode—was a mummy who was treated like a member of the family. And this was a small, tight-knit carnival family, which made her story all the more intriguing.

By the time we got to Marie, one of her caretakers was a man in his fifties named Randy Bridges. Marie's new owner was Captain Boswell's brother, Jim, but Randy was as responsible as anyone for her well-being. He had started working for Captain Boswell's traveling show when he was a teenager. It was one of those classic "run away and join the carnival" stories. You could see the admiration this guy

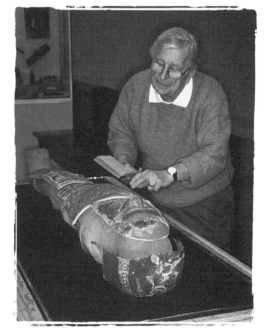

George Sulman and his mummy.

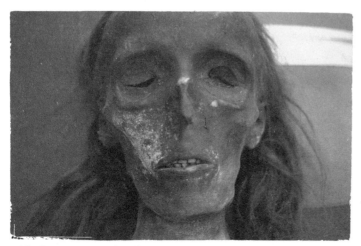

Sideshow star Marie O'Day.

had for Boswell, almost like a son for a father. He wanted to take care of Marie as well as Boswell had—and she was definitely well cared for. Clearly, she had meant a lot to the Captain.

Whereas Hazel Farris, the sideshow mummy from Nashville, had spent time in a garage and was a bit moldy, Marie was stored in her own homemade wood and glass case. Captain Boswell transported her from one location to another in a trailer adorned in big lettering that read, "MARIE O'DAY'S PALACE CAR." He always bragged that she was easily his most popular attraction. During the episode, we hung up the original banner line advertising Marie, and Randy's eyes welled up. It reminded us how alive and vibrant this sideshow culture had once been.

Marie was in spectacular shape. Like most sideshow mummies, she had an almost wooden look to her, which is a sign that she had been embalmed with an arsenic solution. While this type of appearance is characteristic of all sideshow mummies, she was exceptional in this regard. We could tell that Marie had been an attractive woman, a real looker. She had a great story behind her, which made her doubly intriguing, and an interesting pathology

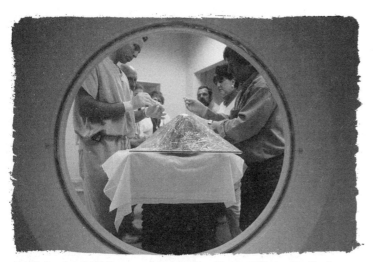

Radiologist Jeff Jones of North Carolina's Wilson Medical Center prepares Marie for a CT-guided needle biopsy of her lung.

to boot. What a wonderful episode this was.

Marie O'Day was supposedly a nightclub singer. The story was that in 1925, her husband—it's always the husband—stabbed her in the back, slit her throat, and threw her in the Great Salt Lake in Utah. Her body washed up on shore twelve years later, and the re-markable condition of her remains was attributed to the lake's high salinity. Like the Egyptians thousands of years ago, whoever concocted the tale knew that salt was an excellent dehydrator and perfect for desiccation.

Marie O'Day's classy means of transportation.

We set up our equipment outside next to the old Palace Car, did our thing, and found no evidence of a slit throat or entry wounds that would suggest she had been stabbed by a knife. We would have seen some type of knife trauma prior to her death. As for the salt mummification, this story seemed highly im-plausible even before we saw Marie. Once we got a look at her, we knew it was untrue. A salt-dried mummy looks much more dessicated than Marie ap-peared. This is one of the reasons we suspected arsenic embalming.

When we looked at X-rays of Marie's organs, they began to confirm our suspi-cions. In fact, we were able to get even more specific in our hypothesis, surmis-ing she had been embalmed shortly after death. Post-mortem decomposition fol-lows a predictable course. The first things

The slow and careful unwrapping process of Marie O'Day.

to go are the intestines (or guts, as we like to say), followed by the rest of the organs in the abdomen, and then the chest. Shadows of Marie's organs were clearly visible in the X-rays, indicating that decomposition had not proceeded very far and the organs were in good shape. Also, in a naturally dried mummy, the eyes typically dry up and are lost, either to insect activity or decomposition. But her left eyelid was open, and we could see remnants of her eye. On Randy's suggestion, we took a tissue sample from her back for analysis, and sure enough, it showed that arsenic was the preservative.

The first things to go are the intestines (or guts, as we like to say), followed by the rest of the organs in the abdomen, and then the chest.

What we found on the conventional X-rays also showed up very well on a CT that was taken by Dr. Jeff Jones at the Wilson Medical Center. A lesion on Marie's lung suggested that she suffered from tuberculosis. We extracted a small tissue sample from her lung for a biopsy, but because the sample was so tiny, the results were inconclusive. Therefore, we took another for DNA analysis. Bingo! It tested positive for TB. This was not surprising.

If you look at the epidemiology of the early 1900s, when Marie appears to have lived, tuberculosis was the leading cause of death in the United States. It was so common that poets romanticized women with the disease. Women dying from tuberculosis were thought to be beautiful—pale, fragile, and breathless.

So what did we learn about Marie? Well, she was not a traumatic murder victim, and she was not mummified through any natural process. Without a doubt, she had been embalmed with arsenic. (We also found the chemical additives mercury and sodium in her system, making her one of the most heavily processed mummies we had ever seen.) We do not know if Marie was really a nightclub singer. We don't even know if her name was Marie. Our best guess is that she was either a homeless person, or that she died in a place where no one could contact her next of kin. Finally, her unclaimed (and now mummified) remains were sold to an entrepreneur, who put her on the sideshow circuit.

One of the cool things about this episode is how much we learned about Marie's longtime owner, Captain Boswell. We discovered him through Gretchen Warden at the Mütter Museum, who communicated regularly with members of the old carnival community. The Captain was one of the last sideshow impresarios around, and Marie was probably the last mummy to be featured in a traveling show. Going into the episode, I think we kind of pictured Boswell as a robust, swaggering caricature of a sideshow operator, but this was not the case.

According to Randy, after eight years and two wars, the Captain had suffered a paralyzing accident while in the navy, jumping from one ship to another, but he was still pretty ornery. If Randy messed up, Boswell would roll up to him in his wheelchair, wrestle him to the ground, and beat on him. Later, when we did a show in Orlando, we met the last surviving sideshow performer who had worked for Captain Boswell, and heard more stories like this. You could tell what a family they were, and how loyal they were to one another.

In the case of the "Mummy in Shades," the mummy literally *was* family. Along with "The Cave Mummies of the Philippines," this was one of my two favorite episodes. As discussed previously, we went to Thailand with the understanding that this former monk, Luang Pho Dang, had been able to meditate for fifteen days—that's two weeks and a day—with no food and no water. Ron and I both teach anatomy and physiology, and while we agreed that it might be possible to last fifteen days without food, surviving that long without water was a stretch.

When we arrived, we were fortunate enough to spend some

Captain Boswell's story was almost as wild as Marie's.

time with Luang Pho Dang's son. He told us his father was a businessman who woke up one day and decided that he was going to dedicate the rest of his life to becoming a Buddhist monk. He had this life, this family, but he just decided to join the monastery, and that was that. Luang Pho Dang was totally committed to his new way of life. Indeed, he wanted to be mummified, thus serving as eternal inspiration for the power of Buddhist teachings.

Luang Pho Dang's son told us his father died in the 1970s and that, yes, he actually could go without food or water for fifteen days. The doctor had told Luang Pho Dang that this was not a good thing to do (in fact, he often ended up in the hospital after these periods of meditation), because it would damage his kidneys and other organs, but he did so anyway. One day, he had a premonition that he was going to die. With the goal of mummifying himself, he began this long process of meditation—again, no food, no water—and this time, he did indeed pass away. We suspect that, due to his debilitated state, someone may have assisted him with a saltwater purge right before death, to help kill whatever bacteria was left in his gut, and dehydrate his tissue even more. As a result, his organs were among the best preserved we've ever seen, especially considering that he was often outside in a humid climate.

So what is it like to have a mummy in the family? We asked Luang Pho Dang's son how he felt about having a father who was a mummy. Although he answered that he was proud of his father and in awe of the powers of meditation, I sensed something in him that told me he felt it was a little bizarre to see his father in this state. There had been so little bodily change that he was still quite recognizable.

Luang Pho Dang's body did not reside in his family's home. He was cared for by his fellow monks at Khunaram Temple. We found it remarkable that the monks were so open to the idea of having a mummified monk whom many of them must have known. It actually worked out well for the monastery, which was located on Koh Samui Island, a popular tourist area. While some monasteries struggle, this one seemed to be doing extremely well. Given the choice between visiting a temple with a mummified monk sitting in the lotus position and one without one, where would you go to leave your offering? The monks do not "market" Luang Pho Dang as such; it just kind of happened that way.

By the way, we doubt that he died in the lotus position—he probably fell over and they put him that way as a tribute to his devotion of Buddhism.

The monks were also surprisingly open to the idea of our investigation. As a rule, we will go only as far as people let us. I can x-ray a mummy without taking it out of its case, if those are the ground rules. But they were very comfortable with us. Before we knew it, they had taken the glass cover off the case, and were encouraging us to interact with the mummy.

Since we didn't speak Thai, we couldn't discuss the concept of meditation with our host, Phra Aduna, without an interpreter. After we finished filming the episode, Master Tui, who spoke fluent English, arrived on the island and explained some basic Buddhist concepts. He took us up to a stream at another nearby monastery, sat us down on some boulders, and talked about Buddhism. What a remarkably gentle belief system. Respect for all life—how can you beat that?

Luang Pho Dang, the Buddhist monk of Koh Samui, Thailand—our "Mummy in Shades."

Sometimes it takes a student body to care for a mummified one. I'm referring to Butch, the mummy at Naperville Central High. As Ron mentioned earlier, you could definitely see this was a potential recipe for disaster—you know, bringing it to football games, putting it in class photos. But to these kids, the mummy was a revered artifact.

As you might expect, our presence was cause for considerable excitement. However, there was also some trepidation on the part of Laura D'Alessandro, the woman who was brought in to serve as the mummy's conservator. We thought we might be able to extract some DNA from a tooth to help confirm its sex and possibly determine the cause of death. But this would involve partially unwrapping the mummy, and Laura was very nervous about us doing that. After careful consideration, she agreed to handle the delicate job herself. We got the tooth and sent it out to a DNA lab. In the meantime, we examined the bones for any disease patterns.

We were not the first scientists to work on Butch. The mummy's primary caretakers at Naperville High were a pair of social studies teachers, Jim Galanis and Tom Henneberry. Some conservation measures had been carried out by Laura at the Oriental Institute in Chicago, and at one point, radiologists at a local clinic had done a CT scan, but they were unable to determine the mummy's sex with these techniques. The best clue that Butch was a male was the cartonage—the mask—which had a man's face painted on it. But when DNA tests on the tooth came back, the results suggested the mummy was a female. Thus, Butch became Butchina.

Ron and I were so impressed by everyone at Naperville High. The amount of time and money that community must have poured into its school is mindboggling. The teachers were fantastic, and the kids were as bright as my best students at Quinnipiac. When we presented our findings about Butch's real sex, everyone seemed even more concerned for the mummy. What better home could you want for something that precious?

Andrew Nelson, an anthropologist, friend, and colleague up in Canada, once heard about an Egyptian mummy that had become available, and he bought it for his lab. It was part of an exhibit that traveled from museum to museum. It ended up with Canada's Department of Education, which wanted to locate an appropriate home for it. Now Andrew's students study it and care for it reverentially; it is housed in a climatically controlled container. Again, you could not ask for a better caretaker.

The most intriguing caretakers of mummies are the collectors, the individuals we sometimes identify as members of the "Mummy Underground."

People collect everything, including some who just happen to collect mummi-fied remains. Perhaps this makes them unusual, but personally I find it eas-ier to understand what moves a mummy collector more than, say, a collector of Pez dispensers. Pez collectors, however, can display their prized possessions to anyone interested in seeing them; mummy collectors cannot. It's a bit taboo in Western society.

Yes, it gets a little strange with these guys sometimes. But that is un-derstandable. Think about it: you own a mummy. What does that mean ex-actly? Where is the receipt? Where are your ownership papers? Can you actually own a dead body, which is what a mummy is? The laws on mummy collecting are a bit fuzzy. Technically, mummies are dead bodies, and the legal system tends to frown on people who collect dead bodies. If you transport one across state lines, a whole new set of laws comes into play. Mummy collectors, therefore, are always apprehensive. They worry about someone suddenly ap-pearing at their door and saying, "You can't own this."

Rather than battling this point out in court, mummy collectors would prefer to maintain a low profile and spend their money on new acquisitions. Thus, they all know each other, but very few (outside the profession) know them. They move among us with complete anonymity, rarely discussing their field of interest and never talking about their collections with an outsider.

There is one fellow out there, a private collector, who has a mummy we were very interested in. Getting to see it became this whole clandestine process, where he would call us from various locations around the country so we could not trace his calls. We could not call him directly. We had to call another num-ber, leave the number where we were, and he would call us back from a phone booth somewhere. He was always moving around. This went on for the better part of a year. He was also talking to Larry Engel, using a different name.

As was often the case, Gretchen Warden connected me with this person. And it took her years of negotiation before I was even allowed to meet him—which is about par for the course with these guys. When I did meet him, I drove to a location and followed him on what seemed like a ten-mile drive to go another two miles. This cloak-and-dagger stuff may sound silly, but I totally understand where he was coming from. He has human remains in his possession that are incredibly valuable, and he does not want to lose them.

This guy is like Superman. He has this incredible secret, but he leads a Clark Kent life. He is not ostentatious in any way, and he does not seem any different than you or me. His house looks like anyone's house. You could pass him on the street or share an elevator with him, and there would be nothing to suggest that he has this unbelievable, diverse collection. Even more impressive than what he owns is what he *knows*—he is a warehouse of knowledge. Interestingly, however, he never allowed any of his mummies to appear on *Mummy Road Show*.

The Mummy Who Out-Acted Lee Majors

Mummy collecting is not for the faint of heart. And the collectors out there are pretty tough to scare. The one story they find positively bone-chilling, however, is the tale of Elmer McCurdy.

Back in the 1970s, ABC was shooting an episode of *The Six Million Dollar Man* using a Long Beach, California, amusement park as a backdrop. There was a mummy called Elmer McCurdy who had been covered with neon-orange paint and placed under a blue light in the fun house. For decades, everyone had assumed he was a dummy—they made fake mummies for sideshows for many years, so this was not uncommon. When the director asked to have Elmer moved for a scene, his right arm broke off and fell on the floor, exposing real bone.

The amusement park employees, cast, and crew were horri-fied, and the authorities, including the LA coroner's office, were summoned. After an autopsy and much publicity, the State of Cal-ifornia issued a death certificate. It was determined that Elmer had been an outlaw who lived and died in Oklahoma. Authorities there, including the governor, requested that Elmer be returned for a proper burial. He went into the ground under a couple of tons of concrete in the Summit View Cemetery in Guthrie, Oklahoma. This put a scare into private collectors, who are afraid that if mummies or mummified human remains are found to be in their possession, they will be dragged into court and forced to bury their collection.

From my experience, mummy collectors are not unlike other mummy caretakers. They treat their mummies with a certain degree of reverence, and in some ways, the mummies receive better care in their hands than they do in some of the situations Ron and I have encountered. Collectors pay a lot for their mummies—$10,000 and more in some cases—so the mummies are more likely to be kept in a controlled environment than those stored in a garage, such as Hazel Farris.

And yet, they live in constant dread of that knock on their door, of that group demanding that their mummies be interred. No one wants to end up on the six o'clock news, or be dragged into court; no one wants to face off against people who find the idea of mummy ownership offensive. These are not only no-win situations in legal terms, but they could also lead to the loss of their collections.

Personally, I think mummy ownership is a gray area. If you own a mummy and aren't breaking a law, why shouldn't you get to keep it and enjoy it? There is no law I am aware of that specifically covers mummy ownership, and the ones on the books that might be applicable seem a bit ambiguous to me. Even more ambiguous is the question of who actually owns human remains, mum-

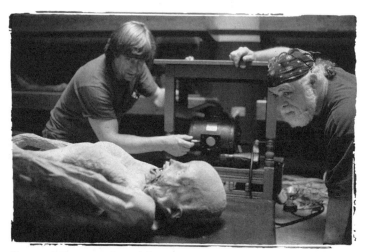

Ron and I set up the Picker Portable at the Museo de las Momias in Guanajuato, Mexico.

mified or otherwise, as well as the artifacts found with them. If an Egyptian mummy is seized from a collector in, say, Texas, does it go into the ground in Texas or does someone in Egypt have the right to ask for it back?

This has come up in relation to museum collections recently, in regard to Native American remains. We worked with Janet Monge, a curator at the University of Pennsylvania, and she was in charge of repatriating Native American remains. Many of the tribes no longer exist, so who should have control of the material? It gets to be a really confusing issue.

When collectors meet with me, they really want my guarantee that I am not going to tell people who they are, where they are, and what exactly I saw. That can be frustrating, because it means I cannot publish the results of the work I do with them. As a member of academia, publishing is obviously one of my goals. So when you're involved in something like this, it has to be done more for the experience. You can learn from their materials, but you just cannot publish what you have learned. Consequently, there is a lot of really valuable knowledge locked inside heads like mine that can never be shared in a formal way. Still, you have to establish these types of relationships or you will never have the opportunity to meet more of these kinds of people, and acquire this important—and unique—information.

The closest we came to publicly dealing with a private collector on *Mummy Road Show* was Dick Horne at the American Dime Museum in Baltimore. He has collected some wonderful sideshow memorabilia, including a Fiji Mermaid and a Devil Child. These are fakes, of course, but Dick and collectors like him are right on the fringe of the folks who own actual mummies. Needless to say, *Mummy Road Show* did not get into any private collections (though we did examine one of Dick's fakes, the fabulous Princess Hana-Luka). Now, if a really extraordinary mummy came on the market, would the Dick Hornes of the world be interested?

That depends. I can think of one he would find intriguing. There is a story of an individual in Oklahoma who, right before his death, claimed to be John Wilkes Booth. In the 1930s, his mummified remains were x-rayed, and a ring reportedly belonging to Booth was said to have been found inside his abdomen. The mummy then disappeared, resurfaced briefly near Philadelphia, and then disappeared again. This is one of those Holy Grail mummies. If it ever came on the market again, I think it would be tough for a collector's interest not to be piqued. And I can't say I would blame him.

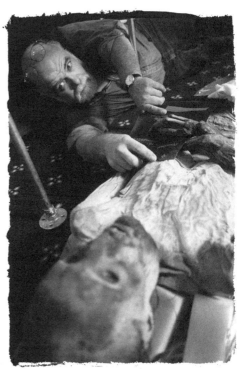

I play contortionist with Hazel Farris.

Not only are there underground collectors of mummies, I think there are underground conservators, too. It only makes sense. If you acquire mummified remains in need of restoration or conservation, where do you turn? If there is a flood or a hurricane or some other disaster that damages your mummy, who do you call? The economics of the situation almost cry out for someone with a specific set of skills and tools, who is willing to work on these mummies and keep his or her mouth shut. Come to think of it, we may even know a guy who would fit this profile. We'll ask him the next time we see him.

I am sure someone will read this and ask, "Isn't it a conflict of interest to work with mummy collectors when they may be doing something that is wrong? Shouldn't you be turning them in instead of safeguarding their identities?" These are fair questions. Initially, my reaction is that it might be similar to reporters protecting their sources—but obviously, it's not quite the same thing.

Ron holds court while I sneak off to check out some artifacts.

The better parallel might be the relationship our colleagues in South America have with the *huaqueros*. They not only know who the looters are, they know who the *best* looters are. They regularly employ them to assist in finding sites, and even in working the sites. If a *huaquero* can make as much money appropriately excavating a site as looting it, why wouldn't he work for you, and why wouldn't you hire him? It may not be the best way to increase the field's knowledge base, but at least you are preserving what history you can and learning as much from it as possible.

Ron makes the point that the collectors we deal with are, in their own way, preserving a piece of our cultural history here in the United States— from the sideshow mummies to the material brought over here during the mummy craze of the Victorian era. It seems weird to say that a preserved human body should be regarded the same way a statue or book or antique from the past is, but in a way, that is what they are. The prices collectors pay seem to me a partial reflection of this.

Indeed, there's a lot of money changing hands in the mummy-collecting subculture. We heard that Hazel Farris was once on the market for as much as $10,000. The truth is that many mummies are priceless. And I am sure there are some great mummies out there that will never surface because of the ambiguity of the laws that apply to owning them. There are books on mummies that catalog what is out there, but you never see most of them because they have been acquired by private collectors. And a few, like Hazel and Elmer McCurdy, are gone forever.

How many mummies *are* out there? It would not be an exaggeration to say that, over the centuries, a million mummies have been uncovered in Egypt alone. They have been dispersed, destroyed, discarded—who knows where they all are now? You have the sideshow mummies and all the mummies of South America. When you account for all the mummies in museums and all of the mummies in private collections, the number is probably staggering.

What is really stunning is the number of mummies owned by museums that the public never sees. Most museums have been receiving mummies from all over the world since the 1800s—more than they know what to do with. In fact, I would not be surprised to find that the vast majority of mummies in the world are rarely gazed upon by human eyes. Ron and I looked at thirty Peruvian mummies in storage at a major university that they probably had not looked at for quite some time, perhaps more than eighty years!

This is hardly unusual, especially if you know a little about how museums work. Consider the number of dinosaur bones and other prehistoric fossils that were dug up and shipped to museums in the 1800s. A lot of those are still encased in plaster, untouched for more than a century. It is no surprise that some of the most important finds in recent years have come not in the field, but as a result of untouched or long-forgotten museum inventory finally being examined. Remember, a typical museum has less than ten percent of what it owns on display at any one time. It can be compared to shopping at a tag sale: You see something that looks neat, you buy it and bring it home, and then you put it away and forget you ever owned it.

Well, the same thing happens with mummies. Museums brought tons of mummies and objects back, and over the years some of them found their way out of the museums and into the hands of private collectors. The collectors we know would be quite happy to relieve a museum of what is essentially excess inventory.

In addition, museums face the challenge of returning all materials in their collections that rightly belong to native people. This is a result of the Native American Grave Protection and Repatriation Act (NAGPRA), which President George H. W. Bush signed in November of 1990. The new law protects burial sites on federal and tribal lands. In addition, it requires federally funded institutions, such as museums, to return cultural items, including artifacts and human remains, to the native tribes from which they had been taken.

Chapter Six

That's Gotta Hurt—
Ron Beckett

One of the advantages of working with mummies (besides the fact they rarely complain) is that you can react honestly when you see something really bad. I don't mean broken-finger bad; I'm talking about axe-to-the-forehead bad. This is not a luxury we have when working with live subjects in a clinical setting. When Jerry and I see a serious injury or advanced disease, we have to keep cool for the patient's sake. But on *Mummy Road Show*, with the cameras rolling, we were encouraged to give the Engel Brothers what they call "reaction shots."

In many cases, these moments really made the shows. We saw some neat stuff, and we saw some unexpected stuff. But what stands out in my memory is when we saw something that made us cringe and say, "Man, that's gotta hurt!" I remember feeling that way after the autopsy of Hazel Farris determined she had pulmonary adhesions. This happens to be within my field of expertise, and I cannot begin to describe how uncomfortable this condition can be. With every breath, the adhesions pull and stretch, causing great pain.

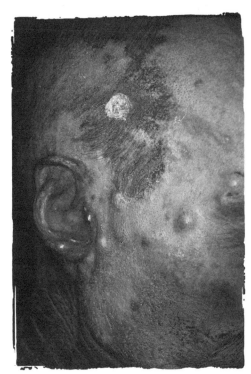

The bumps on Sylvester's head turned out to be steel pellets from a shotgun blast. Man, that's gotta hurt!

Dragon Breath

You do not have to be an expert to imagine the pain caused by the oral pathology we observed in the three years we filmed *Mummy Road Show*. My God! Think of your worst toothache and multiply it by ten. The infections we saw? When they got into the bloodstream, they killed people. Keep in mind that oral hygiene was an oxymoron hundreds of years ago. In fact, one of the revelations we had during the *Road Show* was how often "bad teeth" could actually be considered the "cause of death," up until very recently.

In "Mummy on a Mission," filmed in the small town of Guano in Ecuador, we found evidence of possible jaw cancer related to poor oral hygiene in the remains of Friar Lazzaro. A religious man who went to South America in the 1500s to convert the indigenous Baruja Indians to Christianity, he oversaw the construction of the first Catholic church in the area. His mummified body, discovered after an earthquake in 1949, was put on display in a glass case in Guano's municipal library.

Simply put, the Friar had one of the worst sets of choppers we've ever seen. This wasn't uncommon for his era. His idea of oral hygiene probably started and ended with the occasional toothpick after meals. You could count Friar Lazzaro's teeth on one hand, and the few he had were misaligned because of a deformity

Say cheese! Our friend the Friar had horrible dental problems, and possibly cancer of the mouth.

in his jaw. This meant he experienced pain whenever he took a bite of food.

The Friar also had this incredible consolidation in the jaw area, a condition that could have been caused by an infected salivary gland or mastoiditis (a bacterial infection of the area of the skull behind the ear). Whatever it was, it was localized around the jaw. From the size of his head in general, we thought this must have been a robust guy. But it turned out that most of what looked like fat was a result of this massive swelling around the jaw.

Poor oral hygiene, however, was not the Friar's only health problem. Our X-rays showed that he also had arthritis in the hips and two of his vertebrae were fused together, conditions that were possibly the result of years of horseback riding. There were also signs of a broken rib, perhaps suffered from a fall from a horse.

Friar Lazzaro was a mess from head to toe, including severe hip problems.

The bottom line is he had pain in his back and hips on a regular basis, but this isn't what killed him.

Our investigation ultimately led back to the Friar's mouth. Given his serious dental problems, he would have had a choice of several medical treatments, including applications of burning hot oil, acid, or leeches. We surmised he might have taken a different route. Back then, tobacco was a popular form of painkiller used by the locals. It is very possible that the Friar packed his jaw with tobacco leaves to numb his pain. Over time, the carcinogens from his homemade treatment could have found their way into his system and led to cancer. Still, if it was cancer that he had, it did not kill him. What actually killed him was probably sepsis, a severe illness caused by a toxin-producing bacterial infection. Sepsis would not have been at all uncommon during the Friar's lifetime.

For the Friar, one thing was certain. He must have had one nasty case of halitosis. I am sure if you asked a periodontist today, he would tell you there are a lot of people walking around with bad breath. But there is no way it could compare with the past. People must have had incredibly bad breath. The Friar is a perfect example. When I watch historical movies now and see people locked in long, passionate kisses, my reaction is, "Yeah, right." Unfortunately, we know a little too much about what was going on in medieval mouths.

In "One Tough Cowboy," we introduced the world to Sylvester, a mummy currently residing in a Seattle curiosity shop. This was only our second half-hour episode, so we were still somewhat new to the game. Everything we had heard about Sylvester suggested to us he was a fake. Dr. Arthur Aufderheide, a paleopathologist at the University of Minnesota, believed Sylvester was a fake, a manufactured sideshow mummy. The story was that he had been a bad guy—a card cheat, a gunslinger—and that someone had plugged him on the way out of town. This accounted for a hole in his belly, just the right size for a bullet. He had supposedly crawled off into the desert, where he expired and was naturally mummified by the dry sands of Gila Bend, Arizona—not far from where I grew up!

This was a compelling story. It certainly made Sylvester more marketable as a sideshow attraction than if he had been a teacher or a mailman. But right off the bat, a lot of the evidence did not add up. First of all, getting mummified in the dry desert sand didn't make sense. The human body is 70 or 80 percent water, so a person mummified in the desert would retain only 20 or 30 percent of his body weight. Sylvester was a substantial mummy, weighing well over 100 pounds. Also, he looked as if he were carved out of wood. His skin had the quality of having been shellacked. Sylvester was shiny, almost like one of those antique cigar-store Indians. Nothing we saw in our superficial examination suggested he was the real deal—not the teeth or the hair or even the little bumps on the right side of his head, which we thought was a nice touch on the part of whoever had made Sylvester. If nothing else, the bumps heightened curiosity about Sylvester, stimulating conjecture about what had happened to him.

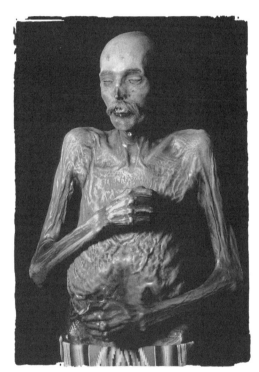

The first tip-off that we might have been mistaken came when we moved Sylvester into position for Jerry's X-ray unit. There was a sticky substance oozing from the body, which is not something you see in fake sideshow mummies. Still, we weren't sure what to make of this. It could have been a sealant painted on the skin to preserve his body.

The clincher about Sylvester's authenticity came when Jerry started x-raying. Lo and behold, we saw a liver. No one would go to the trouble of carving a fake mummy and then putting a liver inside him. So we began to look at the films differently, and that is when we had our eureka moment. Sylvester was real!

It now seemed likely that he had been embalmed. We knew that as a mummy Sylvester had spent several decades with a traveling circus. His caretakers had prepared him to go on the road from one whistle stop to another.

After examining Sylvester, Jerry and I agreed that he was indeed "One Tough Cowboy."

As for those weird little bumps on the side of Sylvester's head, they weren't man-made or caused by a dermatological condition. We had Sylvester

It revealed some amazing information, most notably that each one of those bumps covered a tiny, round piece of metal. Sylvester had taken a shotgun blast to the face! Man, that's gotta hurt!

transported to the University of Washington Medical Center, where they took a CT. It revealed some amazing information, most notably that each one of those bumps covered a tiny, round piece of metal. Sylvester had taken a shotgun blast to the face! Man, that's gotta hurt!

What was amazing is that none of the pellets had penetrated Sylvester's skull. And since the skin had healed around the pellets, this meant that Sylvester had survived the shotgun blast—it could not have been the cause of death. What about the hole in his stomach? We did not detect an exit wound, nor did we find a bullet in Sylvester's body. And it was not for any lack of trying. We searched for that magic bullet like it was the JFK assassination, but found nothing. There was something in his shoulder, but it looked like part of the shotgun blast. So how did the hole get there?

We determined that this was created during the embalming process. We knew at this point from chemical analysis of Sylvester's skin that his amazing condition was the result of an arsenic solution, just like Hazel Farris. The device that likely accounted for the hole in Sylvester's gut was a trocar—a long, beveled metal tube used to deliver the embalming fluid, still used by morticians today.

Once we knew the shotgun blast had not killed Sylvester—you know how we think—we wanted to get a feel for how it went down, and how much it hurt. After the filming of this episode, Jerry took the X-rays over to the medical examiner's office to confirm that these were indeed shotgun pellets, and to gain a little insight into what kind of gun was used, and at what distance. The ME confirmed that Sylvester was hit in the head with a shotgun blast, and estimated from the scatter pattern that he was not shot from that great a distance. It was probably a self-packed buckshot, he observed; there just wasn't enough powder in the shell for a lethal blast. Lucky Sylvester. The ME also said the shot was not lead, because lead would have flattened out from the impact with the skull. Using electron microscopy, it was later determined that the pellets were steel. I'll say it again: Man, that must have hurt!

One final observation from the University of Washington Medical Center added an interesting slant to Sylvester's story. During this era, a doctor could easily have removed the metal shrapnel from his head. This made us wonder: why didn't he seek treatment? Was it because he was on the run? We may never know.

Another eureka moment came during the filming of "Tales from an Italian Crypt," when Jerry and I found a kidney stone in the Italian nobleman we were studying. This thing was as big around as a quarter and had all these spikes coming out of it. Oh man, this was too big to get passed through the ureter, so it probably got stuck in the guy's kidney. He went to his grave with constant, piercing pain. Lopping out the kidney would have been the only way to deal with it back then, which would have been as good as a death sentence.

A close-up of the kidney stone I removed from the mummified nobleman in "Tales from an Italian Crypt."

The amount of pain everyday people endured in past centuries was kind of a revelation for us during *Mummy Road Show*. Intellectually, you know how diseases, injuries, and other chronic conditions could affect people's lives in those days. You also know that medicine was either primitive or nonexistent in many cases, particularly where the poor were concerned. But when you see it with your own eyes—over and over and over again—you do begin to feel their pain a little more personally.

In "Mamma Mia Mummies," we went to Italy to examine the mummies—18 in all—at the Church of the Dead in Urbania. The fact that the bodies were preserved at all was amazing. The climate of Urbania is so moist, not the type of place where you would expect to find mummies. Ultimately, we solved the mystery. The bodies had been wrapped in cloth shrouds and buried in a cemetery with dry limestone soil, a great vehicle for mummification. Limestone can halt decomposition in a number of ways, including its ability to draw humidity out of a closed environment.

Who was responsible for carrying out these 18 funerals? A group called the Brotherhood of the Good Death, which provided burials for the poor, criminals, and people who didn't have families. These were individuals from the bottom rung of society, so we expected to find lots of pathology. And boy, was that the case. Severe scoliosis, kidney stones, bladder stones, arthritis, pneumonia, heart disease—these people were a mess. One of the mummies had a large gash in his chest, supposedly from a stab wound that punctured his heart, but we could not prove this was true.

Interestingly, a physician who treats people living in Urbania today told us there is a lot of correlation between the disease patterns he sees now and the ones we were able to identify, particularly with the older residents. Until the 1980s, people who were born in this valley tended to stay there their entire lives. After 1980, the epidemiology changed as people migrated in and out of the area, and it is now more like the rest of Italy.

In "The Princess Baby," we saw that even children of royal stature were not immune to the hardships of life. This little girl, Anna, was supposed to have been three or four when she died. We determined from looking at her teeth and other developmental structures that she was probably less than two. On her X-rays we looked at the Harris lines near the ends of the bones. When you get really sick, the bones stop growing and a line of calcium forms. The body does not expend energy growing if it needs to address another problem. When you get well (that is, *if* you get well), the bones continue to grow, but that telltale line of calcium remains. This is something we can see with the X-ray. Anna had several of these growth-arrest lines, which suggested she was rarely healthy

These were individuals from the bottom rung of society, so we expected to find lots of pathology. And boy, was that the case. Severe scoliosis, kidney stones, bladder stones, arthritis, pneumonia, heart disease—these people were a mess.

during her short life. She must have had recurring ailments like diarrhea, and probably respiratory problems, too.

When Jerry and I think of the Middle Ages, we think of something glamorous, like Richard Harris in *Camelot*. (For those of you a bit younger, think Heath Ledger in *A Knight's Tale*.) But this period of history could be extremely unpleasant. Anna was the daughter of King Ludwig, who was fighting for control of Bavaria. He was on the march, moving his family around, living in very rough conditions. It's no wonder that she was ill on several occasions in her young life.

Dragging a kid around from battle to battle would be considered child abuse by today's standards (though King Ludwig didn't really have much of a choice). What we encountered in "Mystery of the Masks," by contrast, appeared to be something diabolical. In this episode, we thought we might be looking at the mummified remains of an abused child. We had begun studying these mummies in 1996, before we started on *Mummy Road Show*. An orthopedic surgeon at Yale thought the fracture of this child mummy's forearm was an example of abuse. We had a radiologist, Dr. Tony Bravo, look at it, and he thought it might be a post-mortem injury, a fracture occurring after death.

This is the problem any time you work with mummified remains. If you cannot do an autopsy or question people who had knowledge of the individual's death, no matter how good you are at employing non-destructive methods, there will always be questions that you cannot definitively answer. The process therefore involves a fair amount of interpretation and speculation, so you have to be very careful about jumping to conclusions. What we will do in a case like this is say, "There is a fracture that occurred *around* the time of death." And really, that is all we *can* say. Now, if there is a particular pattern to a fracture, the situation is different. A fracture of the ulna and not the radius, for example, is a defensive fracture—especially when it is more deformed on one side than the other. That would be pretty clear-cut.

This wasn't the case in "Mystery of the Masks." We were looking at a child mummy we had studied three years earlier. Full-body X-rays revealed that she had been a very sick little girl. There was evidence of trauma throughout her torso, and she had the arm fracture. This made us think of child abuse, because this was the sort of injury that may have been caused by a severe rotation of the forearm. We couldn't say for sure, however, because the fracture could also have happened post-mortem.

We took the mummy to the Shoreline Clinic in Connecticut for a CT. X-rays and CT scans—which can detect the location of broken bones, the number of fractures and the extent of healing—are often used to reveal child abuse. The scan of the little girl showed that little if any healing had begun in any of her fractures. This finding more or less ruled out child abuse. Instead, we turned our focus to a nutritional disease like rickets. This made sense because this mummy was uncovered during an excavation across the Nile from the modern city of Luxor, in the area that was once Thebes, a center for healing in ancient Egypt. Perhaps this girl's parents brought her to the city hoping to find a cure for her condition. In any event, she unfortunately suffered a lot in her short life.

Another aspect of life that caused a lot of unpleasantness for the mummies we investigated on *Mummy Road Show* was their working environment. There was no OSHA looking out for common laborers, that's for sure. For example, in "Mummy in Vegas," we originally thought the individual we were studying had worked on the railroad because of his ethnicity—he was Chinese. But the toxicology told a different story. We had a tissue analysis done on his hair, and huge amounts of mercury were found in his system. This guy was actually a miner.

We know this because a common method for extracting gold back in the 1800s involved the use of mercury. This man had been exposed to quite a bit of mercury, probably in vapor form from the mine he worked in, and it almost certainly devastated his central nervous system. We also found a fracture of his humerus, which was never properly set by a doctor. For a guy doing hard labor for a living, that had to hurt like crazy every day. Given the damage to his arm, there was no way he could have wielded a sledgehammer or a pick without an extraordinary amount of discomfort. Based on our conversations with historians and other experts in Nevada, our guess is that he started out as a miner,

We hunker down with village elder Baban Barong and Orlando Abinion of the National Museum in "Cave Mummies of the Philippines."

This is a mummy from one of the hundreds of delicate bundles that Ron and Jerry examined.

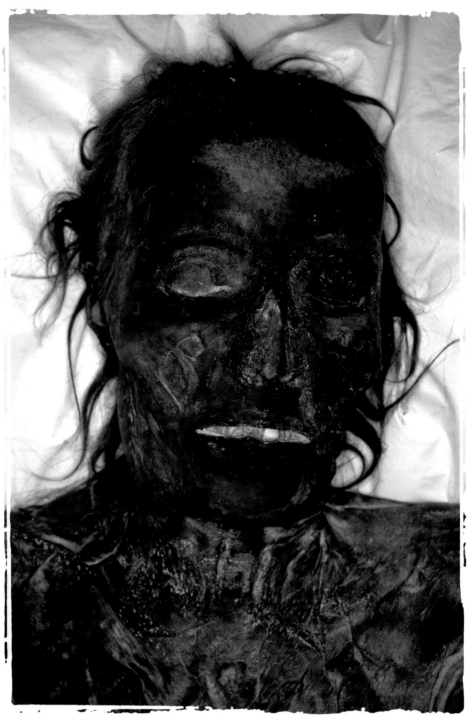

Hazel Farris, a sideshow mummy from Tennessee with a lurid background story.
Hazel is one of our all-time favorites. Her owner allowed us to perform an
autopsy—a rare privilege in the world of mummy investigation. The results
were *very* interesting.

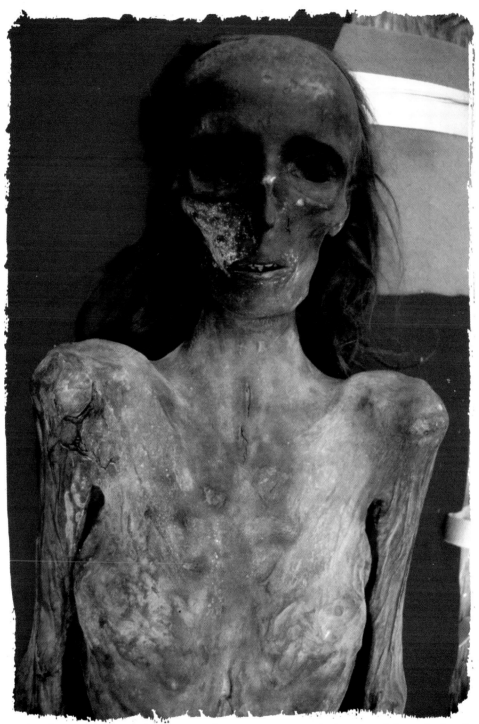

Marie O'Day, a mummy who traveled from town to town in her own deluxe "palace" car (see page 91). The story went that she was a nightclub singer whose husband stabbed her in the back, slit her throat and dumped her body in the Great Salt Lake, where it washed ashore—naturally mummified—a dozen years later.

Luang Pho Dang, a former Buddhist monk in Thailand, whose son is actually still alive. He who would meditate for 15 days at a time without food and water. That white dot on his nose is a lizard egg!

Luang Pho Dang in the flesh. Today his body is cared for by his former fellow monks.

Some mummies were so well-preserved that their eyeballs and most of their teeth remained intact.

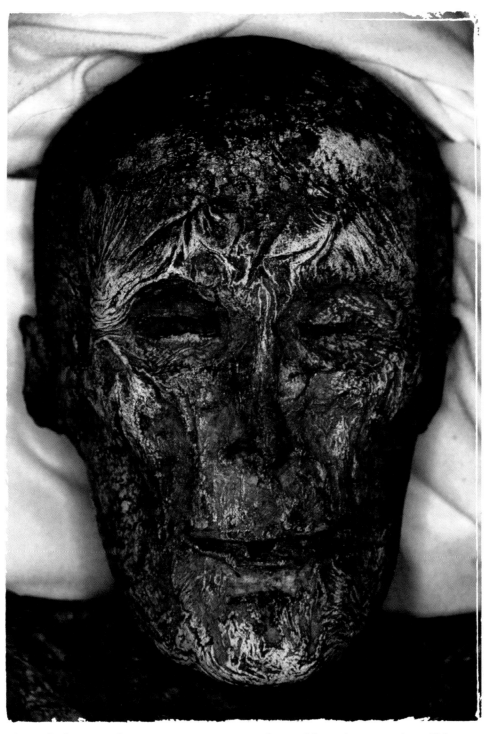

The techniques used to preserve a mummy—along with environmental conditions—determine how it looks as time passes. This mummy was so well preserved that she survived a flood.

Mummies are often found with their mouths wide open. Scary!

These three mummies—each with an intriguing story—lie beneath 1,000-year-old St. Michan's in Ireland. The church's crypt contains hundreds of coffins.

had an accident, and became a cook or laundry man, perhaps in a railroad camp. This poor guy had an incredibly tough life—and he probably never found his "pot of gold," either.

We suspect that some sort of occupational hazard was also responsible for the demise of the individual we examined in "Carnival Mummy." I thought this was one of the high points of the series. Jerry and I were able to show that Andy, our carnival mummy, died from a blow to the chest. After we x-rayed the mummy, we were looking at the films with a radiologist, Dr. George Stanley of the University of Central Florida, and the light went on for all three of us simultaneously. It was a tension pneumothorax.

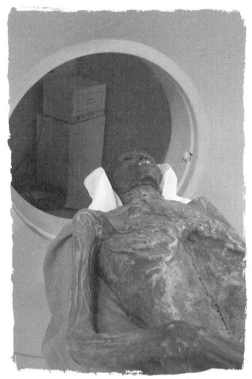

Andy the carnival mummy goes through a CT scan.

This is just a horrible way to die. First, you experience massive trauma to the chest, and then a broken rib that collapses a lung and creates a one-way valve for air to enter the chest cavity from the outside. When you breathe in, you create negative pressure in the chest, and the wound just keeps sucking air in. Everything starts moving around inside—the lungs close down and the heart shifts—and within minutes you are basically dead, unless someone knows exactly what to do. The problem is that only a doctor knows how to treat this injury—a layman would never think to place a "chest tube" where the pressure is building, and then relieve that pressure and allow the lung to re-expand. Anyway, there are nerve endings in the pleura (the delicate membrane that lines the thorax and folds back over the lungs), and when they stretch and pull apart, it's very painful. You can't catch your breath, so your adrenaline starts pumping and your blood chemistry begins to change, which makes your heart work even harder. But the pressure keeps blood from getting back to the heart, and the individual's condition worsens very rapidly.

For this particular case, the quality of the CT and the level of preservation enabled us to confirm the cause of death. We were also able to estimate how long it took him to die from this injury—less than an hour. We could not, however, tell if it was a whack with a baseball bat, an elephant stomp, or something else. He could have fallen on one of those enormous circus tent stakes. That certainly would have caused this type of injury. We do know that, as a carnival worker, it was unlikely that he had access to any kind of health care. His co-workers probably gave him some rum and watched him die. And as we mentioned earlier, he may have been a human blockhead, so they also may have stuck a nail up his nose after he died as a way to memorialize him.

On occasion, we saw the results of medical attention. One of the Kabayan mummies we studied in "Cave Mummies of the Philippines" appeared to be a woman who had died in childbirth, and there was clear evidence that she was given a caesarean section in order to save her baby. This was remarkable—this mummy alone could have been the focus of an entire episode. Jerry was so fascinated I literally had to pull him away from her.

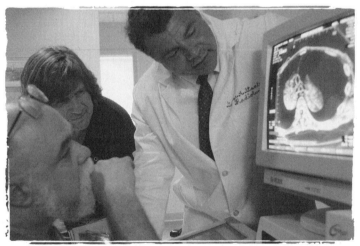

We examine a CT scan of Andy's chest cavity with Dr. George Stanley.

We had one of the villagers with us, Juliet Igloso, a woman who worked in the local museum, and Jerry mentioned his suspicion about the c-section to her. According to her, the evidence of the caesarian birth made sense because there is an old tale that says during ancient times, when a woman was having difficulty giving birth, they got a piece of bamboo, sharpened its edge, and removed the baby through the abdomen. The mother was cared for and tended to until she died. There were no artifacts to put this mummy into a temporal context, but she could have been anywhere from 500 to 1,000 years old. Imagine c-sections being performed in what had to be a Stone Age culture. It just boggles the mind.

In "Holey Mummy," filmed in Cuzco, Peru, we saw wonderful examples of early neurosurgery by the Inca, probably to relieve pressure on the brain. This procedure was often performed on warriors. Blows to the head were an occupational hazard for Incan soldiers, as were the terrible headaches that resulted from battle wounds to their skulls. To fix the damage and reduce the pain, Inca surgeons would remove a small section of the skull, permanently, and the skin would grow back over it. Sometimes the skull would grow back somewhat and cover the hole.

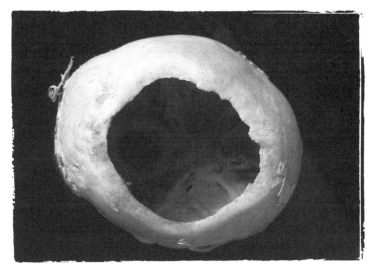

This Inca soldier actually survived one trephination— an ancient cranial surgery—it was the second one that killed him!

The remarkable thing about this procedure, which we know as trephination, was how successful it appears to have been even back 500 years ago when medicine was borderline barbaric. One of the mummies we examined actually went through two trephinations. This was one tough warrior. He got whacked in the head, had surgery to repair the damage, and then later charged back into battle. He wasn't as lucky the second time around. He received a wound near the first one, and his surgeons could only do so much. He died from blood build-up beneath his skull.

The skills of the Inca surgeons were astounding, particularly given their primitive tools. John Verano, professor of anthropology at Tulane and a colleague of ours, has done research on large numbers of trephinated skulls. Get this—the Incans appear to have had a success rate of more than 70 percent.

Neurosurgery in the early twentieth century had about a 20 percent success rate, probably because of infection. During the Inca era, people would undergo the surgery, pack the wound with some leaves or herbs that acted as a natural antibiotic, and go home. Amazing.

You Rang?

Every mummy on display has to have a great story. We found this was true whether it was in a church or a sideshow. The bottom line is that no one is going to pay to look at a mummified university professor who died in his sleep. You've got to make your mummy an intriguing character, such as an outlaw like Sylvester, or put him through some incredible—and awful—experience.

Being buried alive certainly fits the bill for the second criteria. In the 1800s, the fear of this was so great that special coffins were constructed with surface bells that could be rung by the occupants if they happened to be interred prematurely. One of the mummies we studied in "Muchas Mummies" was said to have been hanged, and another buried alive. Everyone has a fear of being buried alive,

Buried alive? The way this mummy's hands were clasped together made us wonder.

and of course, no one wants to be hanged, so everyone could relate to these scary stories—clearly, mummy-story gold.

The mummy who was said to have been buried alive was interesting because her arms were contorted rather than being folded or at her sides, which are typical positions for burial. She also had marks on her face that were not consistent with the frown lines you would expect to see. We had to wonder: Was she trying to claw her way out of the box? *Yikes!* We could not say for sure.

The story of the hanged mummy didn't wash, however. There were no fractures or separations in the cervical vertebrae, and the trachea was intact. Something around the neck had definitely left an impression—the "proof" that supposedly supported the hanging story—but it was not a rope. More likely, it was simply a high collar on a tight-fitting shirt.

The "That's Gotta Hurt" factor varied from episode to episode during the run of *Mummy Road Show*. We were aware that these discoveries made for good television, and probably kept a bunch of people watching who might otherwise have turned away. Jerry and I tried to be cognizant of what worked and what did not from a storytelling standpoint, and the more shows we did, the better we got on camera. Still, we were always careful not to sensationalize something we did not honestly believe was sensational.

A few years before *Mummy Road Show* came along, Jerry and I were working on a mummy from Chile for a documentary. This mummy had a very interesting fracture pattern, which suggested that access to the bone marrow might have been the goal. There was a film crew that was desperately trying to get us to say that this was evidence of cannibalism. But all we could say was that this individual had an interesting fracture. One of the points made elsewhere in the documentary was that women in this culture would eat their babies after they died. The writers maintained it would have made sense for females to cannibalize their dead children—what better way to bring the baby back inside you?

Well, maybe; but we saw no evidence to that effect, and there was no way we were going to say we did. So often in this business, sensationalism is what sells. We were so fortunate working with Engel Brothers Media and the National Geographic Channel, because they always kept it clean. It is what it is, and because the show was of such high quality, in our minds that was sensational enough.

Putting the Hip in Hippie

One of the most interesting aspects of *Mummy Road Show* was how we were able to look back in time and see how people aged. Depending on the context and the culture and the individual's standing in society, the aging process took myriad forms. One thing Jerry and I saw a lot of was joint deterioration. This topic was, literally, a sore spot in my case. During our second season, my right hip was killing me, and I knew it would have to be replaced. Of course, we worked my situation into an episode about mummies with bad hips. Anything for the show.

I was probably predestined to suffer from arthritic conditions because my mother has some pretty serious rheumatoid arthritis. I think I exacerbated the process with the football I played as a teenager, and later, all the sports I played with my kids. And I know that my participation in martial arts worsened it. My right hip was my attack hip, and I would just jam it out there too much. It would inflame, and without good healing, the bones grew and the arthritis set in. To treat it, I opted for a newer technique that is minimally invasive. They cut just one muscle instead of six. I was about the two-hundredth patient to have this surgery, and the recovery was very rapid. Ten days later I was filming a *Mummy Road Show* episode with Ronn Wade at the University of Maryland. I had a cane, but I hardly needed it.

For a while, I simply adapted the position of my leg. If you look at the early shows compared to the later ones, my legs behaved totally differently. I couldn't squat, I couldn't tie my shoes, and it was very difficult to get in and out of a car. In "Mummy in Shades," I was unable to sit in the lotus position for our meditation instruction—I just couldn't do it.

You do not get your hip replaced until it is really painful, but as we filmed each episode and I had to work my way into more and more unusual spaces, I knew the inevitable was coming. Man, it was frustrating being limited like that. It gave me a glimpse of what older people must experience.

I talked the physician into letting me have my hip, which now sits on a shelf at home, next to my grandfather's ashes. It is protected by a little Anubis statue, some good luck elephants, and a porcelain Bruce Lee. I think I'm covered.

Chapter Seven

Grave Matters— Jerry Conlogue

When Ron and I decided to do *Mummy Road Show*, we agreed to have fun with the project, but not to let that obscure the serious science. One of our primary goals was to show people that science could be really exciting. It is not some bizarre, cloistered profession where you are always stuck in a laboratory. Also, unlike Las Vegas, what happens in the lab *doesn't* stay in the lab. What you learn in a controlled environment has genuine relevance

It's a race against time—and the *huaqueros*—at the "Mummy Rescue" site in southern Peru.

in the real world. Lord knows we got our hands dirty on this show, and I speak for both of us when I say we loved every minute of it.

Something else we tried to convey in doing *Mummy Road Show* was how the real world affects the science. You have an unimaginable number of physical and cultural variables—not to mention the human factor—constantly changing the dynamics of a given situation. Bugs and animals you can predict; humans are a bit more difficult.

Thankfully, the one thing we *could* control was the mummies. Mummies are dead, which is nice. They might hold some really cool surprises, but that is why we investigate them. The more twists and turns a particular mummy took us on, the more intrigued we became, which made for a better episode. Looking back, that was one of the more compelling components of the human factor. Occasionally, we had trouble controlling our reaction to the mummies—part of the human condition, I guess. It seems only natural when you see the remains of a child or a dismembered mummy.

When we saw the furrows on the face of the mummy in "Muchas Mummies," it was easy to believe the story that she had been buried alive. We could only imagine what that would have been like, and it kind of made us step back and regroup for a moment. When we determined how young Princess Anna was when she died, and saw what a miserable life she had led, we could not help stepping into the shoes of her father, King Ludwig. He had his battles to fight and his principalities to conquer, but as parents, we knew how deeply the loss of his daughter must have touched him. From what we saw, Ludwig wanted to ensure Anna was well cared for, for eternity.

Were we ever scared by a mummy? No. This was our avocation before it became our vocation for three years on *Mummy Road Show*, so it was unlikely that we would have encountered anything that really freaked us out. Also, remember that I've spent a lot of time with medical examiners, and Ron has seen a lot in his career in health care. But again, it was the humanness of the mummies that sometimes caught us a little off guard. Like when we encountered a child mummy or a mummy with its eyes open and intact.

Even though our goal was to have fun with *Mummy Road Show*, there was also a lot of serious stuff we had to deal with. The real world wasn't just an encroachment; sometimes we had to be proactive in handling challenges that came our way. We had to reach out, negotiate, make uncomfortable deals—whatever it took to get the job done. Not that we fancied ourselves as Indiana Jones types, but there were also times when genuine peril was just around the corner.

In fact, sometimes genuine peril hits you right on the head. Twice.

Filming *Mummy Road Show* had to be worked into our regular, scheduled academic activities. Every summer for the previous four years, I'd spent between eight to twelve weeks in Peru taking X-rays of mummified and skeletal remains for various anthropologists. One of the locations was in the tiny community of El Algarrobal, located in the narrow Osmore River valley near the southern coastal city of Ilo. Groves of olive trees, planted by the Spaniards in the sixteenth century, are located on the southern bank of the meandering river. Sandy hills, some reaching nearly a thousand feet, rise steeply from the northern riverbank and the edges of the groves. Here I worked with Dr. Sonia Gúillen, a Peruvian anthropologist I have known since 1998, at her research facility known as Centro Mallqui.

Sonia is in charge of hundreds of mummies and artifacts from the region representing the remains of the Chiribaya culture that had occupied the valley from 800 to 1350 A.D., long before the Inca Empire was established. The research center was unique because it provided educational opportunities not only for Peruvian anthropology and archaeology students, but also for radiography students from Arkansas State University, who made their connection to Peru through the Bioanthropology Research Institute that Ron and I established at Quinnipiac in 2000. Groups of these American students would spend about a week at the center studying the techniques of mummy excavation, the plants and animals of the region, and, of course, the basics of mummy radiography.

Peruvian anthropologist Sonia Gúillen.

Among the experts who presented lectures to the students was Marvin Allison, a world authority on mummy pathology. Dr. Allison had worked in

Peru and Chile for more than forty years. At the beginning of every season, he would always remind the students that this was a seismically active region—in case of an earthquake, he would tell them, get under a doorway. Every time I heard him say this, I would smile and think, "C'mon, Marvin, there's not going to be an earthquake here. We've come down here for the past four years and nothing ever happens."

Engel Brothers Media had decided that at the end of my stay at Centro Mallqui, the crew would come down with Ron and we would film an episode with Sonia at El Algarrobal. Several days before their arrival, the last group of Arkansas students completed their work and then switched their focus to a little R&R. Marvin and Sonia had departed several days earlier. Following the traditional fiesta we held for each departing group, the students went off to do a little sand surfing on the steep hills that surrounded the valley. About ten of us remained after the students left, and we continued to celebrate. All of a sudden one of the Peruvians stopped talking and looked very attentive. I'm not sure exactly what he said—it may have been *terremoto* (which means earthquake)—but all the Peruvians were suddenly on their feet and running for the arched exit from the terrace area where we were sitting.

Melissa Gibson, a graduate student from Quinnipiac and one of the instructors at the center, was sitting next to me. I think we realized at the same time what was happening, and that someone was missing. Sharlene Walbaum, a psychology professor from Quinnipiac, had skipped the party and was up on the second floor of the center. Shar had accompanied me numerous times to Peru, and was conducting research on expertise related to anthropology and radiology. Melissa and I jumped up and started for the upper level of the complex. By the time we got to the stairs, they were undulating. I had never been in an earthquake, but I thought things were supposed to move from side to side, not up and down. The motion made it very difficult to stay on our feet, but somehow we got into the upper-level kitchen. Thankfully, Shar was there. And that's when I remembered Marvin's advice—needless to say, I wasn't laughing anymore.

The three of us crowded into the kitchen doorway. I remember looking out at the concrete-block wall next to where we were huddled. It was actually swaying in a wavelike motion. At the maximum movement outward, the top blocks suddenly broke free and crashed to the floor. Because the walls weren't reinforced with rebar (steel bars used to provide additional strength to concrete block structures), they were really unstable. With everything so chaotic, it didn't occur to me what was happening above us—that is, until the first block hit me right on my head. I saw part of another block hit Shar's back before a second piece of concrete struck me.

At this point, I truly thought we were all going to die. It wasn't like you read in books or see in the movies—my whole life didn't flash before me. I accepted what I felt was a certain fate. Then it stopped; well, at least the motion subsided. We ran down the broken stairs and out through what had been the front archway of the center. The six-foot concrete lintel that formed the top of the arch was smashed on the ground. We dashed across the road to where the Central Malqui staff members had gathered, and they welcomed us with hugs. Everyone seemed really concerned about me. After a moment I realized why—my head had been cut by the falling concrete, and blood was running down my face.

I was not in any pain thanks to the adrenaline, and incredibly, everyone in our group was present and accounted for. The Arkansas students had survived the quake on a hillside and made their way back into Ilo, where the community cared for them until they could get out of the country and fly home. The research center was completely destroyed. It was a miracle no one was killed. Up the valley, in Moquegua, they weren't nearly as lucky. A lot of people were killed, and 70,000 were left homeless.

All of this happened on a Saturday, and the rest of the Engel Brothers Media crew was not due in until Wednesday. It was very emotional for me, seeing all this destruction in a place I had come to know over the years. The thousand-plus aftershocks did not help, either. We set up a big relief tent, tended to our wounds, and then did what we could to help the community.

A lot of these people had lost everything, and had no shelter. It gets cold in the Peruvian desert at night, so there was a desperate need for blankets.

Eventually, I began thinking about mummies again. I needed to communicate with Ron and Engel Brothers Media, who were due in a matter of days, but there was no phone service in the area. There were, however, a lot of Internet cafés around, and I found one that still had open lines. We e-mailed back and forth, debating whether or not to do the episode. I was against it at first, but eventually we decided to go forward. Even if we hadn't used the footage in the episode, we felt we would get great coverage of what it was really like in the field. And we could also document what had happened to the research center and museum. Fortunately, the storage area—the *depósito*—on the first floor of the museum had been built using more modern techniques, and the mummies it housed were not badly damaged. The second floor of the structure did not fare as well. It collapsed, ruining many irreplaceable textiles.

Just as the earthquake tooketh, however, it gaveth back. The site at El Algarrobal is well known. Indeed, everyone had been aware for a long time that there were mummies there—though no one knew exactly how many tombs there were, or their exact locations. These mummies are buried maybe two meters or more below the surface, so you have to do a lot of digging just to find out whether you have one or not.

After the earthquake, we went to the site and were flabbergasted. What had once been a relatively featureless expanse of sand now looked like the dimpled surface of a golf ball. There were hundreds of depressions, each one created by the shaking of the earth down into the small cavity surrounding the mummy below. This created a tremendous opportunity for us, but also put us in a race against the looters, or *huaqueros*, as they are called down there. Sonia, having returned to Centro Mallqui on Sunday following the quake, wanted to see how many mummies we could excavate before the *huaqueros* moved in and began their activities.

Finding these mummies and looting their tombs had supported *huaqueros* for years, but it was very hard work given the depth at which the burials were located. The looters would go out at night with these long iron poles, which they would work slowly into the dry, sandy earth. If they felt the pole give, this suggested they had hit an air pocket, a tomb. They would come back and dig for the mummy they hoped they would uncover. If the *huaqueros* were fortunate enough to find one, they would take the textiles and artifacts. This time around, I'm happy to report, the scientists won. The *huaqueros* were more concerned with the safety and well-being of their families than the opportunity for looting. It broke our hearts to see the hardships people were enduring, but at least we could take some solace in the fact that we were able to get several mummies out of the ground.

Ron gets down and dirty with the tomb cam in "Mummy Rescue."

Although it wasn't shown in the episode, an aftershock struck while we were conducting our study of a child mummy. We all ran from the building, with Ron being the last of us out. He wanted to save his endoscope. It seems that I had not convinced him how much falling concrete blocks can hurt!

The Buc Stops Here

Before filming our "Pirate Island" episode, *Mummy Road Show* had to jump through a lot of hoops to get the required permits. Even though we were working under the auspices of Peru's National Institute of Culture (INC), access to the island was controlled by the Peruvian navy. At one point, there was a break-in at the INC facility. A dog was killed and a computer was stolen. Peru's bureaucracy is slow like any other country's, and the break-in delayed the permit for a couple of weeks. At least we didn't have to worry about *huaqueros* at the site guarded by the navy.

The island was used as a training area for Peru's version of SEALS. There were a lot of guys with guns on the island, including a contingent surrounding our encampment, ostensibly for our protection.

At night on Pirate Island (officially known as Isla San Lorenzo), the wind would die down and you could hear things from quite a distance. One evening, there was a commotion nearby. We could make out a small boat approaching shore, and we heard the soldiers ordering it to go away. The boat kept coming, and the soldiers became more agitated. We could hear them chambering a round in their rifles, but this still did not deter the boat. Finally, they radioed for "The Jeep"—the only working vehicle on the island. A minute later we could see these two headlights coming across the island, from several miles away. I have no idea what was in this jeep, but when the occupants of the boat saw it, they finally retreated.

Was it a boatload of *huaqueros* that turned tail that night? Or a boatload of buccaneers? I'm afraid that's a military secret.

I touched upon the whole issue of looters in an earlier chapter, and how the situation is not black and white when you see it up close. People have been looting tombs as long as tombs have existed. And regardless of the culture or the country, they are rarely treasure hunters or adventurers. They are simple people scratching out a meager existence on the fringes of society, often just hoping to survive. Sonia Gúillen, who is trying to get her fellow Peruvians to respect their heritage, which notably includes the Inca, and see mummies as a communal asset instead of a "crop," has explained to us the uphill battle she is waging.

Prior to contact with western people, it probably would have been culturally unacceptable for the indigenous people of South America to desecrate a tomb. These were the resting places of their ancestors, which were held in great reverence. When the Spaniards arrived, everything changed. Their message for the next four-plus centuries was that, if you were not a Spaniard—or at least a Catholic—you were basically worthless. That went for the indigenous people living under Spanish rule, and it went for everyone who had been buried before Catholicism came to the New World.

Tombs were looted by the Conquistadors, their occupants' bones scattered on the ground. Mummy bundles were stacked high and torched. It must have been devastating for those early people to watch their revered ancestors and thousands of years of their history wiped out. But as the centuries passed, generation after generation watched this go on, and they became desensitized as a culture. (They also were converted to Catholicism, so they no longer practiced the indigenous religion that had tied them so closely to their ancestors' remains.) With the bond broken, the remains are now viewed as just bundles of bones, there for the taking. If your children have empty stomachs or you lack adequate shelter for your family, would anything stop you from looting a tomb?

What Sonia is trying to do is change the accepted practice of looting by getting Peruvians, especially the younger ones, to appreciate their incredible cultural legacy—not only the Inca, but the many groups that predated the Inca, which were also incredible. In many cases, these are the tombs that are looted. The government simply can't protect every site, and it may be a little more vigilant about Inca sites because of the tourism they generate.

How bad does it get? There is a site known as El Brujo where the archaeological research was financed by a Peruvian banker. You could barely walk around the place without stepping on a human bone; they were scattered everywhere. Once they are removed from the context of their burial setting—taken from their bundles and separated from the artifacts—the bones lose much of their value to the anthropologist. That being said, the bones still have value to scientists like Ron and me, because from them we may find evidence of diseases such as TB, as well as telltale signs of things like head fractures, battle wounds and procedures such as a trephination. But the value of bones at a looted site, in terms of understanding an ancient culture in broader terms, is all but lost.

Bones lie scattered within walking distance of the highway. Who knows what information has been lost?

Say we find a detached arm bone with a large prominence where the muscle was attached. This tells us that we may have an individual who was engaged in something that involved repetitive motion. Without the artifacts or the textiles, we cannot determine much else. Had there been some kind of ocean-going rowing tool with the bone, we would have had a bit of evidence to support the conclusion that this might have been a fisherman. This information, in turn, might help us understand other items in the bundle, or shed some light on the patterns in the accompanying textiles. It might also tell us something about the next bundle, or perhaps even prove to be a key to understanding the nature of the site itself. So every bone that gets crunched under foot represents potentially priceless information, lost forever.

Everything we know about these cultures comes from one of two places: the Spanish chroniclers who observed life in the early stages of conquest, and the tombs. To appreciate just how important the tombs are, you have to know a little something about the chroniclers. They began writing about life in South America towards the end of the sixteenth century, a good fifty years after the Inca last ruled, and society had already undergone cataclysmic change. Plus, they were writing about the vanquished from the perspective of the victor, which further skews the information they were recording. It is like asking the cavalry to write about the culture of the people of the Great Plains. Would you trust what you read?

We find it fascinating, for instance, that nowhere in the Spanish texts is it explained why, for centuries, the indigenous people prepared their dead as they did. Why are there mummies in the first place? That would be nice to know. That information was certainly available to the Spanish chroniclers. Either they did not ask anyone, or they decided it was not worth writing down.

Now you see why what we find in a tomb is extremely important. It is the only shot we have of truly understanding what the lives of these people were like. When we detect evidence of arthritis in a particular region of the skeleton in a group of mummies, we can surmise that they were laborers. If their bundles are undisturbed, we might find artifacts that tell us what kind of labor they were engaged in. And if a site is not looted, the archaeological record and the historical record are more intact. This is so important with a culture like the Inca, which had no written language and thus offered no clues about

These beautiful artifacts excavated by Sonia Gúillen at the El Agarrobal site suggest that this individual may have been a potter.

day-to-day life in writing. There is nothing that tells us, "Today we went fishing and on Tuesday we plan to weave." So we have to try to learn those things from what is written in the bones and the cultural artifacts.

Getting back to El Brujo, the pre-Incan site in northern Peru, looting had obviously gone on here for years. As part of the archaeological excavation,

We were told that some enterprising Peruvian collected all the bones, ground them up into bone meal, and fed them to the pigs. So anyone who ate pork in Peru that particular year might have consumed a bit of ancient calcium.

guards protected the site from additional looting around the clock. We heard years later that there was not a bone to be found on the surface. We were like, "Where the hell did the bones go?" We were told that some enterprising Peruvian collected all the bones, ground them up into bone meal, and fed them to the pigs. So anyone who ate pork in Peru that particular year might have consumed a bit of ancient calcium. Imagine that happening in our country. If they excavate a cemetery to put an interstate through, no one is going to take the bones and grind them up for animal feed. That is how attitudes may vary in different parts of the world.

This kind of drove home the point for us: In many of the places we visited on *Mummy Road Show*, feeding your family is priority number one. If a looted artifact puts bread on the table, then so be it. We met *huaqueros* and saw how they lived. These are not wealthy people. They live in poverty like much of the population in Peru. When we see a beautiful artifact in a private collection, and we suspect that it was looted, the most painful thing is that we know the guy who found it probably made just enough to get his family through the end of the week.

Will the attitudes change if the economics don't? I believe it is possible. How long will it take to change these attitudes? I think back to the societal changes we put into motion during the 1960s, and realize that a generation or more has to pass before any real headway is made. It certainly does not happen overnight, despite what some politicians promise. Ron and I are both old enough to realize this. We recognize that if you have been a *huaquero* for thirty years, then you are going to continue to be one. But, if Sonia can get a few of the kids to appreciate their heritage, things will definitely begin to change. When they grow up, they are hopefully going to say, "No, I don't want to do this, because these are my ancestors, and I want to honor them."

The work Sonia Gúillen is doing with Peru's young people is great, and, as I said, we think that ultimately it will be effective. But what also has to happen for looting to stop is to educate that person at the top of the food chain. If European and American collectors don't put up the $50 or $500 or $5,000 for a black market artifact, there's no money to trickle back down to the source.

Some researchers have been criticized for hiring *huaqueros* to find mummies. How do we feel about that? We have mixed feelings. There is a law in Peru against grave robbing (although there are few resources to enforce it), so on that basis alone we are against anything that would encourage this behavior. Then again, people are eating who otherwise might not, and the sites the *huaqueros* might have looted are being excavated appropriately.

A Peruvian woman rediscovers an ancient weaving technique.

The more interesting subtext here is that locals are being given the opportunity to participate in the study and preservation of their own heritage. This is what interests Sonia and our other friends in Peru, like Jenny Figari de Ruiz. Jenny is directing a project that runs workshops to teach locals ancient weaving techniques. Once they start producing the incredibly beautiful textiles of their ancient ancestors, they can earn a living selling the materials to tourists. Ideally, when ancient sites are found, the Peruvian anthropologists and archaeologists would like to set up a museum right there and show the local people how they can benefit from their history. If that link can be made—and it has in a couple of locations, and it seems to be working— maybe those grave robbers will use their skills as part of legitimate research that benefits their local tourism industry. It really would be no different than what they are doing now, when they hire themselves out to researchers.

A Line in the Sand

Ron and I are naturally curious, and aren't afraid to dabble in the grayer areas of the mummy world. We will not, however, get involved in projects that we believe to be immoral or illegal, such as mummies and artifacts protected by the Native American Graves Repatriation Act, or which hold no promise to advance the study of mummies—no matter how tempting the opportunity might be. Once *Mummy Road Show* began airing, we would occasionally get calls from people whom we suspected were excavating the mummified remains of indigenous Americans.

This probably happens more than anyone knows. Say you are Farmer Bob from southern New Mexico. You are fencing in your three cows, and while digging a posthole, you stumble upon a mummy. You may not be aware that there is a protection act that prevents you from excavating this mummy, so you call an anthropologist you read about in a magazine or find on the Internet—or, in our case, see on TV. Well, the scientist knows the law says you are not allowed to touch this stuff, even if it is on private property.

We definitely draw the line at this point. What is interesting is that we know others who might be willing to work in these situations. You have a lot of young archaeologists and anthropologists out there, just starting their careers, aching to make their mark. They do not always have the perspective needed to make the right choice. They get excited and overzealous, and then find themselves breaking some ethical boundaries. These are the same individuals who may claim someone else's work as their own. I hate to say it, but those people are out there in academia.

I suppose Ron and I are at an advantage in that we think of ourselves more as "consultants" in this field. We are tenured professors, so we do not *have* to publish papers. Rather, we are happy to come down and apply our knowledge and expertise and perspective to a project that has already been started. We typically present papers in an attempt to share with the academic community that which we have learned. This often leads to collaborations and future research that further advances the science of mummy studies. Consequently, when we are invited to get involved with a questionable mummy, the decision comes down to: *Why do it?* Why risk losing our credibility? No matter how intriguing the possibilities may be, we will always say, "No, thank you."

Although the practice of looting is a constant threat to the study of mummies, the fact that it is constant means that it's also predictable, and, perhaps someday, preventable. Another threat, population growth and the urban sprawl that accompanies it, may not be as easy to counteract. People need places to live, and if mummies are buried under land that has been deemed suitable for new housing, those alive almost always win out and those mummies will be lost. In Peru, where Shining Path rebels

The "House of Bundles" mummy room.

A Chiribaya mummy in burial attire.

drove people from the mountains down into the cities, those cities have now begun encroaching on the surrounding countryside. In two episodes we filmed for *Mummy Road Show*, areas slated for development contained important mummy populations. Once again, the race was on!

At Cajamarquilla, an ancient urban center dating back a thousand years, the present expanding community was encroaching on the site. Jenny Figari de Ruiz and Raphael Segura were able to get a permit from the National Institute of Culture (INC) to excavate the specific areas, but it is only a matter of time before the urban sprawl of the twenty-first century overruns the site. Once the community is built, I suppose they could make a case for going back, but you can't dig holes all over the place in a residential area because people will fall into them. Also, with all of the new drainage created by this population, a lot of water will be seeping into the ground, and the mummies will probably be destroyed.

At another location with extensive burials, known as Tupac Amaru, Guillermo Cock received an emergency permit from the INC to excavate the site. The residents there actually wanted to save the mummies, which made for an interesting story. Guillermo—known far and wide as Willie—wanted to excavate. Although he received the necessary INC permit, he could not get the money he needed from the government. So, this poor community—I still do not know how they managed it—raised enough funds to support the excavation in what is the community's soccer field, and to preserve the mummies recovered. In this case, the race was against both time and dwindling funds. They both ran out quickly, but Willie and his team, including many locals from the shantytown, were able to eventually rescue nearly 1,300 mummies.

By the way, many of these mummies were in bundles, and some of these were big bundles. Everyone has to store their mummies somewhere, and in Peru they do it in depósitos. Willie lives in one of the suburbs of Lima, and we visited him there. Willie took us over to this average-looking house, no different than other houses on the block. I believe he rented it to use as an office—and a place to store mummies. The space certainly did not go to waste. Inside there were mummies stuffed everywhere: on shelves, in boxes, and in all the rooms and hallways in all three stories of the house, It was phenomenal! He had all the Tupac Amaru mummies in there. That's Willie for you—he wants to save every mummy he can, and he'll take matters into his own hands if necessary.

The good news in Peru and other South American countries is that there are a lot of mummies still left. In many regions, like the southern coastal areas of Peru, anything that goes into the ground is going to be a mummy pretty quickly. This is due to the extreme arid environment of the Atacama desert, plus the saltiness of the sand, especially with desiccating bodies early in the decomposition process. There are probably more mummies in this part of the world than there are even in Egypt. This is an exciting thought, because in places like El Algarrobal in the Osmore River valley, where we worked with Sonia Gúillen, the looting activity has probably only made a slight dent in the enormous amount of mummies buried there. There are probably many thousands of mummies still waiting to be discovered—an entire culture under the sand—that appear to be members of a peaceful

The sawed-off legs of a thief? Or the result of a one-size-fits-all coffin?

farming society. Thus far, not one of over two hundred mummies I have radiographed in this region has showed any signs of violence. Not only have the X-rays suggested they weren't clashing with other societies, but it also appears they never even fought amongst themselves. This may seem like boring stuff to the layman, but to someone in our line of work, it is an important idea. For some reason, for centuries, these people just did not fight. I for one would like to know why. The more mummies we study at El Algarrobal, the more detailed the answer to this question will be.

Money Matters

One of the pleasures of doing *Mummy Road Show* was that we were not involved in any of the financial negotiating, which, unfortunately, is often part of mummy research. Sometimes museums or other organizations charge a fee to investigate their mummies and other artifacts, and rightfully so. Some are ridiculously priced, some are priced appropriately, and each has a unique set of restrictions. The Engel Brothers staff did a great job taking care of all the little details, so we were able to focus on the work and not sweat the finances or the politics. We worked until someone said *Stop*, and then we let the producers sweat the details. When they said *Work*, we continued what we were doing.

There was only one instance during the three years we did the show when we found ourselves in the middle of a financial dispute.

Chapter Eight

All These Years Mummy's Been Faking It—Ron Beckett

People have been manufacturing fake mummies as long as there have been suckers willing to pay for them—or pay to *see* them. From the Egyptian "collectors" of the Victorian era who hoped to sell mummies to eager Western tourists, to the small-town sideshow hawkers of the twentieth century, to people making some very strange and elaborate counterfeits out there today, fakes have been part of the mummy world for more than a century. And needless to say,

Whether it was a fake or the real deal, Jerry and I were always willing to make house calls on *Mummy Road Show*. Unfortunately, our truck did not always get us there.

we've definitely come across our share of phonies.

Then there are the mummies who are genuine, but whose stories simply aren't. We tend to encounter these types of "fakes" more often. We don't see our role in these cases as debunkers, by the way. We like to think that, through good science, we can paint a clearer picture of how these individuals lived and died. Sometimes the real story isn't as sexy as the one their owners have been telling, and in these cases we have to accentuate the positive aspects of what we've found. As Jerry likes to say, we take a situation that may be bitter tasting to the mummy caretaker and find ways to make it sweeter.

It is human nature to be a little rattled when you find out your mummy's story isn't entirely true, but most people get over the disappointment and focus on all that we *do* find. *Okay, maybe your guy wasn't shot through the heart, but look at all this fantastic new information you never knew about.* I think, deep down, people feel good when they get cold, hard facts about their mummies. It makes the mummies seem more real. Another way we avoid a big letdown is to make sure a mummy's owner gets copies of all the imaging and testing we do, and anything else we can offer that might bolster the exhibit.

You really have to approach the process this way. It takes a lot of trust for owners to let us work on their mummies, especially if they do have an intriguing story. They need to believe that we will enhance their ownership, because if we are viewed as "mummy busters," then we simply would not be welcome.

In "One Tough Cowboy," which was the second episode that Jerry and I did, we examined Sylvester, supposedly a card cheat who had crawled off and died in the desert. At first, Jerry and I were almost certain that he was a fake, but he turned out to be the real deal. We discovered some great stuff about Sylvester, including the fact that he had taken a shotgun blast to the face. As much information as we added to Sylvester's story, however, we did manage to determine that he had not been shot in the abdomen, which had been a central part of his tale. In fact, we suspected the hole in his gut disproved another crucial part of Sylvester's story—that he had been mummified naturally in the dry desert sands of the American southwest. It turned out that he had actually been embalmed, and the hole was where the arsenic solution that had preserved him so gloriously had been pumped in.

Sylvester's owner, Joe James, had inherited the mummy from his father. Thousands have had the privilege of viewing Sylvester at the Ye Olde Curiosity

Shop in Seattle. (They don't charge a fee for a glimpse of him, but many of the people who walk into the store to see Sylvester walk out after having bought something.) Even though Joe had wholeheartedly believed the legend of Sylvester, he seemed to like the idea that we uncovered the truth, which helped breathe a little life back into this unknown character. However, his son and daughter-in-law, Andy and Tammy, did not. The couple now runs the shop, and I sensed a lot of concern on their part that Sylvester's story had changed. Sylvester was their future inheritance, and they must have wondered whether people would still come to see him now that his tale was slightly less fantastic. Personally, I doubt that this will have any effect on whether people come to their curiosity shop to see Sylvester. He is an incredible, exquisite modern mummy, and well worth the trip.

Nevertheless, Jerry and I are very sensitive when presenting the truth about a mummy. We've come to realize that, in some cases, telling someone their mummy is not exactly what they say it is—or believe it is—might be like telling them, "Sorry, this isn't your father; it's someone else."

News of the truth of a mummy can really stir up emotions, especially if the caretaker has come to view the mummy as an extended family member. So, in the case of Sylvester, for instance, what we tried to say was, "Here are the possibilities: I don't think that hole in his stomach is really a bullet wound. It just doesn't look like a bullet wound. It could have been where an embalming tool went in. But how fascinating was that shotgun blast to the face!"

We have not been back to see Sylvester, so we do not know if they have changed his story for the public. Our policy is to present our data to the owner, and if we end up debunking a myth, it is not our call whether that myth changes or not, because it is not our mummy. We are particularly sensitive to this policy when our findings may have an economic repercussion. This was a consideration in "Luck of the Mummies," the episode about the church in Dublin that housed four mummies with very intriguing stories. Beneath St. Michan's, which is almost one thousand years old, is an amazing network of crypts that houses hundreds of coffins. The bodies are well protected from the

The mummies at St. Michan's Church in Dublin.

elements, and the limestone walls remove a lot of the moisture from the air—a perfect place to make mummies.

The four mummies we investigated were said to be a crusader, a nun who lived to be 122 years old, a thief whose hand had been cut off, and an unknown woman. These individuals—billed as the "Big Four"—were housed in the same crypt, which is open to the public. People make donations when they view the mummies, and it is said that rubbing the mummies brings you good luck. The people at St. Michan's were wonderful. They even let me play their famous pipe organ, where Handel was said to have practiced his *Messiah* before performing it in the area. We knew that the mummies were a source of revenue for the parish, and we were mindful of this throughout our investigation.

We x-rayed and scoped all four mummies, and transported the thief to a hospital to get a CT scan. We were curious about some objects in his abdomen that could have been shrapnel. It turns out they were probably tiny rocks, which oddly got in there post-mortem. As for his hand, it was definitely severed cleanly, indicating that he probably lost it *after* he died. We didn't think this was done as punishment, which was the story that had long been circulated about this person. Given that his feet were sawed off so he could fit in the coffin, it is just as likely that his hand was removed and sold to a medical student. Early medical schools often needed specimens for study, and they weren't always picky about how they acquired them.

It was interesting to see that the feet and legs of the next mummy, the crusader, had been removed, too, and placed in the coffin with him. Because he was a large individual, we surmised he simply did not fit into the one-size-fits-all coffins of the Middle Ages. It was not uncommon back then for a body to be crammed into a coffin too small for it. What we did not expect to find was that the feet and legs were much smaller, proportionally, when compared to his hands. As we looked closer, we also found that he had an extra pair of knees. (And, no, he didn't have four legs.) When Jerry's X-ray showed two spines, it was clear we were dealing with two corpses here—or at least one corpse on top of another partial one.

Of course, there was one big question we couldn't help but ask: Was he (or they) really a crusader? When crusaders returned from the Middle East and died, their legs were crossed when they were buried. This mummy's legs were crossed, which was probably how the story originated. But we noticed that his pelvis had split apart at some point, and whoever put the pieces back together had crossed the legs. This did not preclude him from being a crusader, but it didn't prove anything, either. The definitive answer came courtesy of a fabric sample I found in his chest cavity. I was able to remove it with the endoscope, and then sent it to be carbon-dated, along with a sample of lung tissue. The numbers that came back said he had lived two hundred years *after* the Crusades.

It was interesting to see that the feet and legs of the next mummy, the crusader, had been removed, too, and placed in the coffin with him. Because he was a large individual, we surmised he simply did not fit into the one-size-fits-all coffins of the Middle Ages.

Of the last two mummies we examined, both females, the unknown woman did not tell us much. The details of her story were sketchy, and other than the fact she had bad teeth, there was not much we could add. The nun was a different story. She had a multitude of bumps on her arm, which sort of gave her the appearance of great age. We were not sure how her legend originated, but from what I saw inside her skull, she was not close to 122 years old. In fact, from the sutures in the skull plates, she appeared to be no older than sixty, and perhaps as young as her thirties. As for being a nun—we weren't able to determine this.

The bumps turned out to be very interesting. When we took a closer look, we noticed the nun had two elbows on her left arm, which suggested that this was a mix-n-match mummy. We asked our friend, pathologist Larry Cartmell, about the bumps, and he thought they were calcium deposits, probably a result of chronic kidney failure. He also added that the arm did not belong to the nun, because its owner would have had these awful bumps all over his or her body. You could see how this condition would have made someone believe this was an incredibly old woman, but the evidence pointed to someone much younger.

I wonder if St. Michan's would have let us investigate the mummies had they known the outcome of our study. Not only did the church lose stories about a crusader and a 122-year-old nun, but the canon—a wonderful guy with whom we lifted many a Guinness—broke his foot descending into the crypt while we were filming. When we told him his mummies did not appear to be what the church said they were, he was kind of quiet and contemplative. I don't think they changed the stories. And that's fine, because the mummies are very important to this church. The parish does noble work for the community, and the mummies bring in a little money from the curious, which is put to good use. Also, we could not say with absolute certainty that the tall fellow was *not* a crusader, or that the woman with the bumps was *not* a nun.

Is it wrong to mislead people by embellishing stories to make their mummies more interesting? I don't think it is, particularly in this context. Besides, when people go to see these mummies, there is a sort of secret voyeuristic thrill that no one wants to admit. So there is some element of self-deception on the part of the patrons.

This dynamic has always been present at American sideshows. When you ventured into that tent, deep down you knew what you were

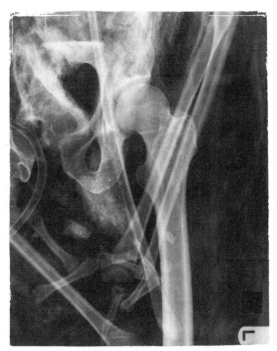

Jerry's X-ray reveals the mismatched limbs in a coffin at St. Michan's.

seeing, or the elaborate story you were hearing, probably was not real. But it is a lot easier to say, *Hey, let's go look at this mummy of an outlaw*, than, *Hey, let's go gawk at some dead people*. Well, the same holds true for the tourists who go to see the four mummies in Dublin. They say, *Hey, let's go visit historic St. Michan's—and by the way, there's a tour there where we can see the mummified remains of a crusader in the catacombs*. There is definitely a titillation factor involved.

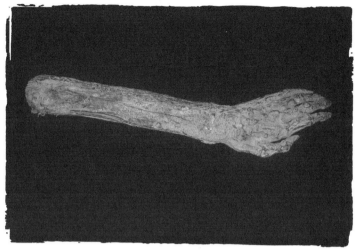

The mysterious arm of the nun in St. Michan's. Turns out it wasn't even hers!

The truth is that you never know how someone will react to the news that their mummy's story does not hold up under scientific scrutiny. In the case of Hazel Farris, who was supposedly a notorious murdering prostitute, we found evidence to suggest that parts of her story were completely untrue, and we suspected that perhaps none of it was true. Hazel's owner seemed almost relieved that she was not the woman her lurid story suggested.

Sometimes we *are* specifically asked to play the "mummy-buster" role. Just because a mummy looks real, and you can trace its ownership back a century, there are still no guarantees that it is the genuine article. There are some really neat fake mummies that have been in circulation since the late nineteenth and early twentieth century—not just alleged mummified people, but cats and birds and monkeys—that were made for the purposes of exhibition and thus sold as real mummies. Obviously, Jerry and I can detect these with our diagnostic imaging methods.

Still, there are some great fakes out there.

In fact, there were many times on *Mummy Road Show* when we took a superficial look at a mummy and thought it had to be a fake, only to find out it was real. And, to the credit of the individuals who crafted the fakes, there were many times that we thought the phonies were real.

These guys were artisans. They were good at what they did because there was a steady demand for the items they sold—and fakes were not just restricted to mummies. In the heyday of the American sideshow, the Nelson Supply Company offered all sorts of interesting things. I've thumbed through the old catalog, and you could buy anything—two-headed goats, mermaids, even a nine-foot Amazon princess—and they were all really well done. Some came complete with banners and other materials. You'd open up the crate and it was all there. You were ready to go—a turnkey operation.

We know from our investigations that the people who built mummies had a good idea of what they were doing. They liked to use bones from cows, and they left the bones exposed in strategic places to make their mummies appear genuine. They really looked authentic. They didn't anticipate the X-ray and endoscope, however. When we x-ray the Nelson mummies, it is easy to spot the wood and nails used to construct the armature. But to the naked eye, they are right on the money.

The Amazon Princess was clearly one of Nelson Supply's. For an extra hundred bucks, you could also get her mummified baby. We examined an Amazon Princess at the American Dime Museum in Baltimore for an episode called "Faking It." It was exquisitely done. I know Jerry likes the fakes as much as the real ones for this reason—the craftsmanship was truly extraordinary. These specimens also gave him an opportunity to get some unusual X-ray images.

The beautiful Amazonian Princess from the episode "Faking It."

Whoever built the Amazon Princess was not only familiar with human anatomy; they also clearly had studied mummies from South America. One giveaway that this was a fake was that she was not in the fetal position (which would have been problematic from a display standpoint). Other than that, however, she appeared quite authentic. There was even an opening in the skin that revealed the rib cage. We were able to determine that these were cow bones, but nonetheless, they had the desired effect.

As mentioned earlier, one of the regrets we had while doing *Mummy Road Show* was that we so rarely had the opportunity to follow up on the discoveries we'd made. This goes for the real mummies, of course, but it also goes for some of the fakes. I would have loved to figure out the Amazon Princess. I bet we could have brought in a hardware expert who could have given us a time frame based on the nails and screws we found. And with a little legwork, we might have even discovered the identity of the individual who made the mummy.

Jerry works on the nine-foot Amazon Princess at the Dime Museum in Baltimore.

Giving a New Meaning to "Jar Head"

People are still out there creating new mummified remains, and selling them to collectors. They know that an X-ray or CT scan will reveal the truth, but they also know that it is unlikely a

Graham Hamrick's sulfur-based embalming solution. With a label like this, you know it's got to be good.

collector will bring his new purchase to a radiology lab. It's not the same as having a used car checked out by your mechanic.

Jerry met a collector a few years back who had paid $10,000 for a mummified head in a bell jar. The story behind the head—yeah, there's always a story—was that this fellow was a thief in India who had been stealing horses from the British. They caught him, lopped his head off, stuck it in this jar, and put it on display at the officers' club. The skin on this mummy looked 100 percent authentic, and because the head was sealed in this bell jar, you could not examine it without destroying the value of the piece. What a great story, and what a great mummy! Only there was one catch. When Jerry x-rayed the head, he didn't see a skull—this was a fake. This collector was not a happy camper. He was out a great deal of money.

What Jerry did see was a skull-like shadow that seemed strangely familiar. He went to a hobby shop on a hunch and purchased a Lindbergh skull. Lindbergh is the company that made those great artificial models, the Visible Man and the Visible Woman, dating back to the 1950s. Jerry assembled the Lindbergh skull and x-rayed it. Lo and behold, the shadow inside the mummy was a perfect match.

The fear that mummy owners have of being exposed has given mummy counterfeiters an opportunity to dupe collectors out of thousands of dollars.

If you buy a fake, who do you go to? The police? The FBI? These collectors would never risk that. The legal ramifications of admitting to owning a human body are not worth it.

If someone tried to sell *us* a fake, we would be hard to fool. That is because we have the equipment and the experience to spot a counterfeit. Imagine that you are an individual without access to this expertise, however. And imagine that you are being offered a mummy that is extraordinary, something that is, for all practical purposes, priceless. Assuming you had the money in the first place, you would probably be tempted to purchase it for your collection, even at the risk of losing your investment down the road if it turned out to be a fake.

This is not to say that mummy collectors are unsophisticated. On the contrary, they tend to be very sophisticated. But the guys who make the fakes are pretty sophisticated, too. They have gotten good enough to fool advanced collectors, and even museums.

Duping eager buyers with phony mummies isn't a new phenomenon. This has been going on for a long time. In ancient Egypt, fake animal mummies were sold as offerings at temples. Fast forward to the nineteenth century, when travelers would visit Egypt intent on bringing back a mummy. The mummy dealers were in the business of making as much money as possible, so naturally, they tried to offer the most intriguing specimens. If that meant getting a little creative, then so be it. As far as they were concerned, the tourists were getting what they wanted, and the dealers were getting what they wanted. A perfect example of this was the supposed mummified baboon in the episode, "Egypt, California Style," which turned out to be a vase.

In this episode, we initially were called in to look at a mummy in the Rosicrucian Egyptian Museum in San José, California. The museum had purchased an ancient Egyptian coffin from the Neiman Marcus Christmas catalog, which always has some wild, unique items for sale. In this case, an Egyptologist at the museum saw some markings that indicated it might be a royal coffin, so she raised the necessary funds and purchased it. It was guaranteed to be authentic, and the price was reasonable, so what did they have to lose?

I'm going in! We work on a mummy at the Rosicrucian Egyptian Museum in San Jose with curator Lisa Schwappach-Shiriff. This mummy was purchased from Neiman Marcus!

When the coffin arrived, the shippers mentioned that there was something rattling around inside. The museum staff opened it up, and there it was, a "free" mummy, like a stick of gum in a pack of baseball cards. And this is great— the movers actually offered to throw it out for them! Fortunately, the folks at the museum said no. As it turned out, the mummy and the coffin were the real deal.

Of course, in the end, we also examined the supposed baboon mummy, and discovered the truth about it. To their credit, the staff took it very well and seemed genuinely pleased to know what they had and didn't have. That little baboon has become quite famous since our visit, and probably draws more visitors than it did before everyone knew what it really was.

Skin Deep

We found the spookiest examples of mummies that were not exactly mummies in Italy, at the anatomy museum of the University of Modena Reggio Emilia, for the "Involuntary Mummies" episode. The museum was created in 1815 by Duke Francesco IV, who was installed by Napoleon and who presided over the city for several decades. Though he ruled with an iron-clad fist, the duke was also paternalistic. He wanted to take care of everyday people who had no one to look after their bodies after they died. Because he was in charge of the prisons, anyone who died there became his property. This is how he was able to obtain so many bodies for his displays.

The duke also had education in mind when he founded his museum, which was among his crowning public achievements. It was similar in concept to the Mütter in Philadelphia, in that it was a center of learning for doctors and medical students. Some of the displays are just wild, including a pair of mummified individuals who were supposedly the duke's African servants: a woman from Namibia and a man from Ethiopia. We had never seen anything like these two mummies—if that's what they really were. They had a much more lifelike appearance than your typical mummy, enhanced by eerie glass (or perhaps porcelain) eyes that had been added postmortem.

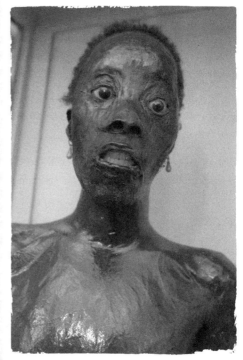

Duke Francesco lived in an era when there was a great fascination with all things Egyptian. Napoleon had brought back countless artifacts from his time there, causing a sensation throughout Europe. Clearly, the duke seemed intrigued by the idea of mummifying bodies, and set out to improve on what the Egyptians had done. During this show, in addition to the two African servants, we examined three female mummies done

Before becoming a mummy, this Ethiopian man was a servant to the Italian Duke Francesco.

during the 1830s and '40s. They were preserved exquisitely, each one better than the one preceding it. Their skin was smooth and well conditioned, and their bone structure and internal organs remained intact. In many ways, the duke had indeed surpassed the Egyptians in quality of preservation.

In the case of the Africans, Duke Francesco's aim appeared to be different. They looked more like anatomy learning tools than mummified bodies. The skin was peeled off the bodies and preserved separately, then sewn back on some kind of lifelike framework. Their skeletons, meanwhile, were on display next to them. Or were they? We x-rayed the female, and as advertised, she had no skeleton. She had been stuffed with gesso. Was her skeleton the one hanging next to her? We couldn't say for sure without DNA testing, which in this case wasn't a feasible option.

When we x-rayed the body of the male, we were astonished to see that his skin had been placed back on his own skeleton. When we x-rayed his head, however, the skull was missing! The skeleton next to this individual obviously was not his, and after a closer look, we determined that the missing skull was not the one perched atop the skeleton. So where did this fellow's head go? No one knows. The only clue to go on was a story that the duke had murdered his servant in a jealous rage after discovering he was having an affair with his wife. Great story, but it did not help us much. Although I suppose it could account for the fact his skull was nowhere to be found.

Were these two African servants actually mummies? I suppose it depends on your definition. Jerry and I discussed this very question, and agreed that—although they were marvelously preserved—they probably fell under the definition of taxidermy, not mummification.

In "Mummy Mismatch," we saw a wonderful mummy at the Wayne County Historical Museum in Richmond, Indiana. This place was fantastic. It had everything from automobile and piano exhibits, to a Barbie collection, to a display on the Wright brothers, who were from Richmond. And the museum owes it all to Julia Garr, a wealthy town resident who was way ahead of her time. She traveled around the globe a century or so ago picking up terrific souvenirs and bringing them back to Indiana.

This might have been the oldest Egyptian mummy we'd ever seen. Mrs. Garr was told this woman had been a priestess who stood at least six feet tall. This would have been incredible for ancient Egypt, especially for a woman. Average heights back then were five-two for females and five-six for males. (Today, we have the potential to grow taller because we meet our nutritional needs much more completely and generally live healthier lives.) In ancient Egypt, someone as tall as this mummy probably would have owed their extreme height to some sort of disease, possibly either Marfan syndrome (a condition that affects connective tissue and causes body parts to become elongated) or a form of gigantism. We know that these conditions were present in some of the royal Egyptian families, but no physical evidence had ever been found. Needless to say, this was potentially a huge discovery.

Jerry and I are not Egyptologists, but we knew the moment we saw this particular mummy that the parts did not fit. It almost looked as if someone had gone through a mummy dealer's merchandise cafeteria-style to put this mummy together. You know, "That's a nice-looking wrap job, so I'll take that. There's a lovely mask, I'll take that. Oh, and put it in that coffin. It matches the rugs." We have seen pictures from the 1800s that show lines of coffins and mummies, and collections of heads and hands, all laid out for sale, so we know this went on.

The mummy itself was obviously too large for the box selected for it. In fact, the wrappings around the feet seemed to have been whittled down at

It almost looked as if someone had gone through a mummy dealer's merchandise cafeteria-style to put this mummy together. You know, "That's a nice-looking wrap job, so I'll take that. There's a lovely mask, I'll take that."

the toes so the coffin could be closed. Now, it *is* possible for a mummy's feet to be worn down to dust over thousands of years by the bottom of the box, but this appeared to be a pretty clean slice.

Also, we learned that the cartonage came from a different period than the wrapping style. So everything was authentic, but the whole presentation was just wrong. This, by the way, is one of the things that makes the study of Egyptian artifacts and mummies so difficult to pinpoint once they have come out of the ground. Because of the rush to market and sell them, they often lost much of the context in which they were buried and discovered. It would not be unusual to find a male mummy in a coffin identifying its occupant as a princess, because someone somewhere along the way thought the princess did not look as good in the box (more on this in a moment).

Jan Livingston, the museum's director, had a lot of questions about this mummy. It had actually been x-rayed back in 1974 when a college student named Mark Millis became very interested in it. He remembered being mesmerized by this mummy as a kid, and he wanted to learn everything he could about it. One really cool thing about this episode was that Mark joined our investigation, some thirty years after he helped launch the original study. It was great because he was still so eager and curious about the mummy.

Among the things we wanted to find out were the mummy's sex and disease history, which would provide clues about this person's real story. As I mentioned, we were told this mummy was a woman, but we didn't know if we could be sure of anything in this case. The skull would reveal a lot. Male and female skulls dif-

The Amazonian Princess in "Faking It."

fer in several important areas, including the brow ridge at the forehead and the mastoid bones on either side. Just by examining the skull as I held it in my hands, I thought there was something fishy. The brow ridge seemed too pronounced for a woman. Jerry reserved judgment until he did an X-ray. The images confirmed our suspicions: this was no priestess. Well, at least not the head of a priestess. It was definitely male.

We also used the skull X-rays to determine if this person had suffered from Marfan syndrome. The answer was no. The skull was simply too thin. Marfan syndrome causes the bones in the skull to be extremely thick. The X-rays also showed that this person had not been a giant. Gigantism affects the ethmoid bone in the skull known as the sella tursica. When we saw that this area had maintained its regular shape and form, we concluded that this skull belonged to a normal-sized man.

Given our findings, we wanted to make sense out of the rest of this mummy. As we viewed the X-rays of the torso, arms, and legs, we noticed how scrambled the body was. This was probably a result of the coffin being moved from one location to another.

Larry Cartmell, Jerry, and I perform an autopsy on Hazel Farris.

And why were the mummy's hands and feet abnormally large? We deduced they were from some other body. Again, this was a practice typical of Egyptian mummy dealers a century ago. But we did check the hips and legs just to make sure. Very large people put more stress on their lower body. Because of this, there are telltale signs of change you'll find in the hips and leg bones that take place over the course of a lifetime. We didn't see any of these.

So when it was all said and done, we pieced together a very interesting picture of this mummy for Jan, Mark, and the Wayne County Museum. While it wasn't a priestess or giantess (as I mentioned, the head was from a "he" and the body was from someone who stood no taller than five-nine), the mummy did have an incredible history. We might have changed the story a bit, but we also added so much compelling new information to it. Priestess or not, this mummy is still the most popular attraction in Richmond.

That says something very interesting about the public's fascination with mummies, and confirms what we suspected all along. Sure, people probably care whether a mummy is the real deal or a phony, but at the end of the day, they care about a great story even more.

Chapter Nine

That's a Wrap—
Jerry Conlogue

There's a little ham in everyone, and Ron and I are no exception. I would be lying if I said we didn't enjoy being on camera for *Mummy Road Show*. We knew going in that we would be expected to move the stories along with our commentary, and to convey our sense of shock or wonder or excitement, depending on the moment. This prospect was a bit daunting at first, but after we got a couple of shows under our belts, we

Ron and I walk like an Egyptian for "Egypt, California Style."

felt totally natural. When the cameras started rolling, Ron and I pretended we were standing in front of a classroom full of students. That being said, by no means were we polished TV commentators when the series ended—often one of us would say something brilliant and Larry Engel would ask if we could repeat it after he switched camera angles or adjusted the lighting. More often than not, we forgot what the hell we had just said.

One aspect of *Mummy Road Show* we did not anticipate was getting involved in the reenactments. When the idea was first proposed to us, we
weren't sure how to react. Ron did a lot of acting in high school, and he loves
to sing and perform, so he seemed a lot more excited than me. And National
Geographic Channel wasn't sure, either. We did not do a tremendous number
of reenactments, but where it helped the viewer to step back in time I think
they worked very well. The reenactments also helped us understand the stories on a more personal basis.

The first reenactment in which we participated occurred on the Hazel
Farris episode, "Unwanted Mummy." It was almost like a mini-movie. I was
asked to play the sheriff who wrestled with Hazel and was killed. Three
deputies had already attempted to subdue Hazel, and all three had met their
demise. Now it was my turn. The camera started rolling; I drew my pistol, and
started to enter the house. The next thing I knew, everyone was falling down
laughing. Apparently, I have a lot to learn about the handling of firearms. I
was going in with my gun held high, *Hawaii Five-0* style. All that was missing was me uttering, "Book 'em, Dano." I still take crap for that one. But
through the magic of editing, it turned out great.

There were a couple of other memorable reenactments. In the "Death in a
Bog" episode, we traveled to Holland to examine a pair of 2,000-year-old individuals that had been unearthed in 1904 by a local peat cutter. Like many European bog mummies, they may have been victims of ritual murder. So for the
reenactment portion of the program, Ron and I got to participate in a mock
human sacrifice. How many people can make that claim? (Or would want to?)

Reenactments were sometimes done at a house I was renting, partly because there was always a lot of space, and partly because there was an out-
building where decomposing material could be stored. This place also had a
church day-care center bordering one side of the property. Twice a day parents would come by to drop off or pick up their kids. For the reenactment in
"Mummy Mismatch," I played the embalmer dressed in faux leopard skin with
a plaster-of-Paris Anubis head on my shoulders, and Ron was the corpse. We
had him laid out on a table, feigning to scramble his brain, a standard proce-

dure in Egyptian mummification. When we rolled him over, we made it look like some snotty stuff was hanging out of his head. Fortunately, no parents happened onto us. I'm sure they might have suspected the worst, and a raid by local police would not have been out of the question. Of course, we were cracking up the whole time, but once again, the final edit turned out really well. It was a very effective way of demonstrating that the mummy-making process could get pretty messy.

The dressing-up part of the reenactments was a blast. We wore some elaborate costumes in the "Mummy Mismatch" episode, and got a lot of interesting feedback about it. It kind of started the cult of Ron and Jerry, which continues to this day. Ron gets letters from all the young women (at least, they *claim* to be young women). What do I get? Well, one guy wanted to be my slave and do unusual things to my feet. Enough said.

Of Mice and an Enterprising Man

One of our favorite characters from the world of embalming (how often do you get to say that?) is Graham Hamrick, a farmer who lived in the town of Philippi, West Virginia, at the turn of the last century. His work was included in "Homemade Mummies." Hamrick may have been concerned with the fact that morticians were harmed by arsenic embalming, or he may just have been looking for a better preservation method. Either way, he decided to make up his own brew, based on a sulfur solution.

Hamrick tested it on a variety of subjects, starting with raw meat and ending up with bodies from a nearby mental institution, and eventually patented the solution. As for the mummies of the mental patients, they ended up on the carnival circuit. Although arsenic still did a better job, the mummies he created were incredible. They survived a flood, which tells you something right there.

Hamrick was a good thinker. He developed an idea, followed it through, and, for a time, his solution was a popular embalming fluid. This guy also was a real character. He would preserve a hunk of meat with his solution and then feed it to people without telling them what it was. Then he would say, "By the way, this meat has been out there on the shelf for five days." I'm sure some of them turned green.

When the town of Philippi heard *Mummy Road Show* would be filming there, people started coming out of the woodwork, bringing their Graham Hamrick memorabilia. One guy showed us a jug of Hamrick solution with about ninety dead mice inside. He claimed they had crawled in there and were preserved. The story just didn't add up, and you can see during the episode we weren't buying it. Mummy research brings out some interesting characters.

Our most memorable reenactment experience, if that is what you would call it, came during the filming of "Carnival Mummy." We were investigating what looked to be the mummified remains of a sideshow "blockhead" at the Ripley's Believe It Or Not! Museum in Orlando, and took a side trip to Gibsonton, the winter home of sideshow performers, near Sarasota. We went specifically to visit Ward Hall, who had been "King of the Sideshow" for more than forty years. The idea was to get a better feel for the kind of culture this mummy might have been a part of. While we were there, Molotov Malcontent stopped by. He's a knife thrower, and, well, one thing led to another, and Ron and I each took a turn standing up against the board as Molotov hurled knives at us.

It was actually my idea. The night before in Orlando, I'd heard that we were going to meet Molotov and his wife (who was his usual target). I thought it would be really exciting to take her place. Everyone said, No, this is crazy, you *don't* want to do this. Once we got to Sarasota, Larry set up the camera to get a target's-eye view of Molotov's work. Between the fact that he did not hit the equipment and the fact that his wife was alive and un-maimed, Ron and I agreed to try it. I mean, if he didn't stick his wife, he probably wasn't going to stick me.

I have to say, it was really kind of exhilarating. You can see in the episode, he came extremely close with the knives. I was standing there in a shirt and tie, with my arms raised over my head, still half expecting that some clever trick was involved. But no, those knives came in hard, and went right into the board. If you think the guy isn't throwing them, or there is some kind of illusion involved, forget it—it was the real deal.

Ron and I agreed afterwards that the experience was worth the risk. We needed to immerse ourselves in that culture to understand the trust and ca-maraderie these carnival folks must have had. They probably thought we were nuts for going through with it, but they also appreciated the fact that we were kind of in awe of their talents. I know that for Ron, who had his son, Paul, with him, it was kind of an emotional thing to see this different, but very real, expression of family. Everyone we met had an unusual skill or phys-ical condition that they were willing to take out on the road to earn a buck or two, and it made for an interesting culture with very strong bonds. It re-ally helped us understand on a much deeper level what we were seeing when we analyzed the Ripley's mummy.

Where did we draw the line on immersion? I can only speak for myself, but I once met an old carnival sword swallower, Red Stuart, who tried to con-vince me he could teach me how to swallow a sword in twenty minutes. I sim-ply couldn't digest that one. (Sorry, I couldn't resist that line.)

Ron and I met some truly fascinating people as the series progressed, and we were also able to involve people we had known prior to the series, particularly in the scientific realm. These relationships were crucial to the success of *Mummy Road Show*, and I maintain that they added immeasurably to the knowledge base that other mummy researchers can work from.

Ronn Wade is an excellent example. He actually had been an embalmer at one point, and provided us with terrific insight into problems we might never have solved on our own. In the "Carnival Mummy" episode, for instance, we e-mailed him a digital photo of the stitching pattern the embalmer had used to close the body. Ronn determined, in real time, that this individual had been worked on no earlier than 1918 and no later than 1926.

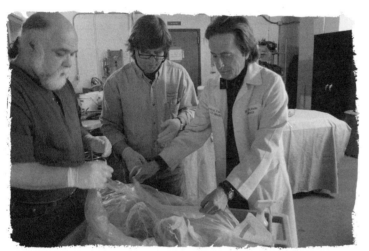

Ronn Wade shows us some of his treasures during "Medical Mummies."

Ronn is the director of the State Anatomy Board for Maryland and keeper of the cadavers at the University of Maryland School of Medicine—he is in charge of body donations. He has a tremendous respect for the science and study of human anatomy in all its variations. The collection of medical mummies there is fantastic. And his plastination lab is unbelievable. This is where the future is going for anatomic specimens. Ronn removes a body part and replaces the water in the tissue (about 80 percent) with silicone. He then performs these elaborate prosections, which display particular parts of anatomy; theoretically, they could last forever. The idea is that you don't need as many bodies to study anatomy. When he shows you this collection, he is like an entomologist with his prized butterflies, except he is pulling out hearts and kidneys and other organs.

Ronn is creative, on the cutting edge, and a really good scientist. On *Mummy Road Show*, we did not always have time to let relationships develop, so we were careful not to go into new situations like bulls in a china shop. This definitely helped in Ronn's case, because I know he was sizing us up when we first approached him. When you hear "TV," you almost have to be a little suspicious. I think we became good friends, despite his early concerns.

Because personalities so often come into play when you collaborate closely with people in our field, Ron and I have been fairly selective about the people we choose to work with. Once in a while on *Mummy Road Show*, we found ourselves paired up with individuals who were, shall we say, without naming names, a little uptight. On the other hand, we had the opportunity to meet some really laid-back people, as well.

In this regard, Bernardo Arriaza, an expert on Chilean Chinchorro mummies, was about the nicest guy you would ever hope to meet. Ron and I had both heard of him before actually encountering him at a paleopathology conference. From that meeting, he invited us down to Latin America to do some work for a documentary on the Chinchorro mummies aired by the Discovery Channel. This was a spectacular opportunity—these mummies are among the oldest in the world. Our next encounter was on *Mummy Road Show*, for the "Mummy in Vegas" episode. We were working with Jennifer Thompson, an anthropologist at the University of Nevada Las Vegas. Since Bernardo was also on the faculty at the university, it was only natural to get him involved.

Bernardo is a world-class scientist. He is intelligent, relaxed, and has a wonderful sense of humor. And he is a good friend. In the mummy community, these qualities are very much interrelated. Friendships often blossom from the kind of mutual respect we share for one another. Bernardo cannot do what we do, and he respects not only that we *can* do it, but *how* we do it. And we cannot do what *he* does. We do not have his background or his deep understanding of the history in the study of the Chinchorro population or mummification techniques. It makes for a great friendship and a pleasant, ego-free working relationship.

The same can definitely be said of Bernardo's peer in the field of bioarchaeology, and our dear friend, Sonia Gúillen. The first time I worked in Peru, in the summer of 1997, she was among a group of anthropologists who had heard about the X-ray field lab I'd set up on the north coast of Peru. It was a tradition on Sundays for the scientists working at various sites around the country to visit one another, and she became intrigued by the possibilities of my portable X-ray unit. In 1998, she called me and asked if I would come down and set up a lab in Leymebamba, where I could work on the mummies in the cloud forest. She told me to get a ticket to Lima, and that she would meet me at the Miami airport.

This was the beginning of quite an adventure. I had never seen Sonia, so I had no idea what she looked like. I got to the terminal in Miami, found my connecting flight, and proceeded to ask every Peruvian in the terminal if her name was Sonia Gúillen. We finally found each other and flew down to Peru. This was the year of *El Niño*, so things were a bit different than when I had last been there. For some reason, it did not really register with me that El Niño would have a long-lasting impact on the weather.

Sonia Gúillen and I work on a site in Peru.

It was raining when we got to Lima, but I had no idea what was happening with the weather in the rest of the country. During the next two days we hung out at her house, had a birthday party for her brother, and spent a day looking for a dog for her to adopt. The Peruvian evening news broadcast scenes of the devastation caused by the recent storms, but it still didn't register with me that we would be in the thick of it. I was so anxious to get out in the field that I never really considered what the conditions might be like, and Sonia did not offer up much in this regard.

On the third day, about a dozen of us piled into a rented bus. Among the passengers was an architect, who was going to submit a design for the museum Sonia was planning, and a number of women, who assisted in curating the textiles and artifacts found with the mummies. There were also two drivers. They planned to take turns driving and, with the exception of stops for fuel and restroom breaks, the trip would be continuous. Thus began a forty-five-hour journey over washed-out roads, open fields, and every other manner of waterlogged terrain. It was surreal.

Finally, we arrived at Leymebamba and I set up my X-ray lab. We used our time well and did some great work on pre-Incan mummies known as the Chachapoya, but soon it was time to go home. The night before the planned departure, there was a birthday party for Sonia. It seemed the entire village turned out for the festivities, which began around ten P.M. I don't remember much in the way of food, but there was plenty to drink. By two in the morning we were playing musical chairs; by four, the dancing had really started; and an hour later, it was time to get on the bus for the return trip. Sonia and most of the others stayed to continue their work, so I set out with the architect and one of the curatorial assistants. There is a point I feel I should make here: my Spanish was limited, and few people spoke any English, including my travel mates. Our goal was to get to Lima, where I would catch a flight back to the United States.

Easier said than done. The first seventy miles (one hundred kilometers) of the road out of Leymebamba was remarkable even when it wasn't raining. Barely two small car-widths wide, it was cut into the steep mountainous terrain that forms the cloud-forest region of northeastern Peru. There were precipitous drops, some hundreds of feet, into the Utcubamba River. The torrential rains had loosened the dirt supporting the rocks that formed the sheer wall of the roadway opposite the river. The resulting rock and mudslides oozed across the road and fell into the raging river below. We hadn't traveled more than an hour when our path was partially blocked by a rockslide. Unfortunately, the Utcubamba was undercutting the opposite side of the road.

The driver and his alternate urged us to help them clear the way before the slide or the river ended our journey. They also reminded us of an important point: If the way forward was cut off, the road wasn't large enough to turn the van around, and a slide behind us would prevent our return to the village. In other words, we would be in serious trouble if we couldn't get through. We scrambled out of the van and began removing as many rocks as possible while keeping an eye out for further slides. After about fifteen minutes, the driver decided he had to make an attempt before the churning river collapsed what road there was left. He had us run ahead to the clear section of road and we watched him gun the van across the path we had cleared. In the pouring rain, we applauded his driving skills and piled back in to continue the trip.

We were so impressed by the driver's skill and nerve that it seemed to us we had somehow made it through the worst. We passed scenes of destruction I never could have imagined. After about thirty hours of continuous driving, we approached the northern city of Trujillo, about three hundred miles from Lima. The city had an airport, so I knew I could catch a plane to Lima. Just outside of Trujillo, the Pan American highway was blocked. People packed the roadway. Enterprising Peruvians had set up stands to sell food and beverages to the crowds. Our driver stuck his head out the window to inquire what was happening. The Chicama River had taken out the bridge, which meant our journey by van was over.

As I mentioned, my Spanish was not great, and I may not have totally understood what everyone was saying. The architect suggested we wade across the river, and for some reason I agreed. We joined a large group of people who had started walking through a cane field to the Chicama. I placed my duffel bag on my head and joined the mass of humanity as they walked into the field. In no time, the water was up around my knees, and the footing became unstable. After about an hour of slogging through this mess, we arrived at the river's edge. I swear it was a mile wide. Did I mention I cannot swim? Did I mention that I have a fear of water?

The farther out we walked, the higher the water got. It came up to my waist, then my chest, then my chin. I could really feel the current pulling me.

Fifty yards in front of me it looked even stronger. Just ahead of me two people dropped beneath the surface, popped up, and then disappeared. I never saw them again. I just kept moving.

Somehow, I made it to the other side. Little did I know that my misery was far from over. The water was filthy with lots of foul stuff, including sewage, and I'm sure I smelled pretty awful. I'm also sure I swallowed a few mouthfuls of the Chicama during my crossing. I barely made my flight in Trujillo, and had nothing to change into. In Lima, I had fifteen minutes to make the connection, so there was no way I could change there. I really smelled bad by this time. The poor people sitting next to me—they must have wondered who the hell had let me on the plane. When I reached Kennedy Airport, I had to face the customs agent waiting to greet me on my return to the U.S. By now, I was beyond ripe. The agent took one look at me and asked

Crossing the Chicama River. My adventure was only beginning.

what the hell had happened to me. I attempted to explain my adventure, but I'm sure my odor made my story far less interesting. He motioned me on, and I somehow staggered into the shuttle van for the ride back to Connecticut. That night I woke up and found I couldn't swallow. Apparently, organisms in the water I ingested during my crossing of the Chicama had caused my throat to swell (a condition known as pharyngitis). I needed medication to treat the problem. The sum effect of my trip had taken its toll, but I knew I would live to tell my tale again and again.

Despite these travel woes, I jumped at the chance to work with Sonia Gúillen again. The next opportunity came when she invited Ron and I to contribute to a documentary called *Desert Mummies of Peru* about the Chiribaya. This was the first time we worked with the Engel Brothers, and it was the

Sonia Gúillen studies a Chiribaya mummy.

genesis of *Mummy Road Show*. During that trip, by the way, a "recent" mummy was discovered in the area where we were working. He was wearing jeans, had a cracked skull, and was buried in the sand. Sonia asked us when we were leaving, and said she would not report the discovery to the police until after we had departed. That way we would not get embroiled in any local drama. Little gestures like that go a long way.

Having colleagues like Sonia Gúillen in this business is critically important to the work we do. When we go to another country, we have no way of understanding the spirituality and culture of the indigenous people beyond a superficial level. Yet, to get good work done—and to understand the work we do—you need that kind of connection. Sonia feels the same way. At the beginning of each field season, for instance, she would take us through this ritual of paying homage to the dead, and also asking their permission to begin work. That is an important lesson that I have learned.

As an anthropologist, Sonia is smart, sensitive, and tireless. She doesn't just excavate for science; she excavates for spiritual reasons, and you can feel her love and connection to those ancestors. Not only does she care for them, she talks to them. If she is working on a baby, she talks to it tenderly, just like a mother would speak to her own child. One of the reasons we get along so well with Sonia is that she appreciates the respect we have for the cultures we study. We also share the same work ethic and dedication to science.

Another kindred spirit, albeit in a totally different sense, is Neil Haskell. We first worked with Neil during the filming of "Mummy Mismatch" at the Wayne County Historical Museum in Richmond, Indiana. He is one of only eight forensic entomologists in the world, and he is a real character. Neil's specialty is describing the succession of insects that feed on dead bodies so he can determine time of death. He brought us out to what he affectionately calls his "body farm," where he has dead pigs put out in different conditions so he can monitor the insect activity on their corpses. When we were there, two grad students had just finished an experiment where they had dressed dead pigs up and assaulted them to mimic a particular double-murder crime scene. All for the purpose of learning, of course.

Neil is an enormous guy, the kind of person you would like to have on your rugby team. He collects big things, too. He had several military-surplus vehicles in a barn, which was very intriguing to us. Neil would have let us drive the tank had it not been raining so hard. It was probably lucky the weather did not cooperate, because I have a feeling we would have taken out a building or two.

Neil shares with us a love of making science come alive. The more creative—or disgusting—the better. His enthusiasm is infectious, which is obvious when you see his students running around the farm collecting jars of insect-laden pig slime. Add the fact that his lab is actually in the basement of his mother's house, and you know you are dealing with a unique individual. Personally, I liked the fact that he was more focused on the science than on the gadgetry. Like me, he had learned how to make old technology do new things.

Larry Cartmell is another down-to-earth scientist who is brilliant at what he does. He is from Ada, Oklahoma, which is not exactly a hotbed for mummies. He works as a pathologist at a hospital, but at every opportunity, Larry is off to Egypt, working

> **He brought us out to what he affectionately calls his "body farm," where he has dead pigs put out in different conditions so he can monitor the insect activity on their corpses. When we were there, two grad students had just finished an experiment where they had dressed dead pigs up and assaulted them to mimic a particular double-murder crime scene**

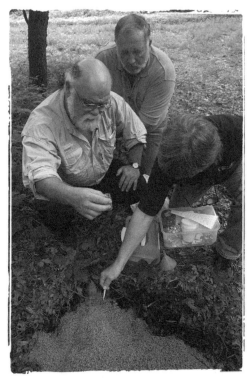

Down on the (body) farm with Neil Haskell.

with the best people in the field of physical anthropology, doing autopsies on mummified remains. It is not only his passion; it has made him one of the most highly respected people out there.

When we received permission to do an autopsy on Hazel Farris before her owner had her cremated, there was no question that we wanted Larry on the case. He helped us learn so much about her, confirming many of the findings we had developed through X-ray and endoscopy, and discovering some other things that added to her story. Larry was great about conducting the autopsy at a pace that allowed us to check every detail from our own analysis. At one point, he stopped everything so I could remove the heart and lungs and x-ray them individually. This led to a more detailed analysis of the blood clots we saw on her lungs. A lot of guys would have whipped through the job in fifteen minutes, but Larry was not interested in making this *The Larry Cartmell Show*. It was *The Ron and Jerry and Larry and Hazel Show*, and he was definitely on the same page with us in that respect.

During the three years *Mummy Road Show* was on the air, we sent Larry tissue samples I don't know how many times. He did great work coming up with tough answers. I feel good that Larry has also benefited from the work we did together on the show. He says he has gained a greater understanding of how certain diseases present themselves in mummified remains. There is a certain poetry to the fact that we are still collaborating on a lung specimen we took from Hazel during the autopsy, exploring some new techniques to analyze it. That means Hazel is still "alive," even though *Mummy Road Show* is off the air.

Heart and Soul

The level of dedication we see in our business is sometimes astounding. People make incredible sacrifices for the sake of knowledge and discovery. This is especially true of the scientists who live in the country where they work. They are preserving their own history, and feel a connection that is difficult for an outsider to fully comprehend. A guy like Willie Cock makes it a little easier.

No one is more selfless or sincere about what he does than Willie. He cares so much for the ancient populations he studies, he is so proud of the work he does, and he is just wonderful to his grad students. As a result, he has taken more mummies out of the ground than he knows what to do with, all in an effort to save them. Man, does he have his hands full.

In the spring of 2004, I saw Willie at a paleopathology meeting in Tampa, and he told me he'd suffered a massive heart attack. He had lost a tremendous amount of weight, had given up smoking and drinking (two things I couldn't imagine Willie ever giving up), and he is still devoting himself to this project in Peru. There he was, encouraging his graduate students who were going to present their papers, and presenting a paper himself. Amazing.

One of the most interesting people we met while working on *Mummy Road Show* was Barry Anderson. Barry was introduced to us as a fan of the program when we were in Orlando at the Ripley's Museum. He is the guy who creates wax figures and puts together the dioramas you see when you visit one of the Ripley's locations. Ron and I visited Barry's shop and saw all sorts of great stuff. There was Jack Nicholson and his famous "Here's Johnny!" scene from *The Shining*, and you would swear you were looking at the real thing.

Barry is a gifted and talented artist. He's like a big kid. He loves what he does, and he does it well. And he is an expert at making mummies. His mummies look so real, in fact, that we later had him make an Inca mummy for the show, which we took down to Peru. It is now in Sonia Gúillen's museum in Leymebamba, and no one can tell it's a fake.

One of the kindest compliments we have ever received for the work we did on the *Road Show* was from Gino Fornaciari, a pathologist who described what we did as "precious." For someone whose native tongue is Italian, this word in English has great meaning to him. He meant it was unique, one-of-a-kind. From then on, whenever we found something unique, one of us would say "precious" with an Italian accent. Ron actually was better at this than I was. Larry Engel finally ordered us to stop.

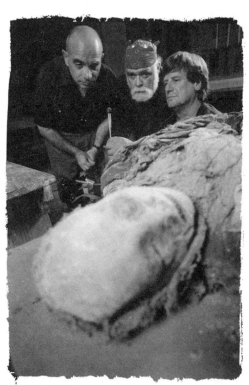

Gino Fornaciari consults on "Tales from an Italian Crypt."

Gino is a very humble guy. Everyone loves him. He was another person who had no issues with ego. He was easy to work to with, and he was genuinely impressed with our approach to the dead. As a pathologist, he does a lot of autopsies, but he was amazed at how much information we could ascertain with non-destructive techniques. We liked and respected him so much that when we wrote up the research on the Italian nobleman from "Tales from an Italian Crypt," we made him first author. He could not believe it—he was absolutely floored. Since then, Gino and his colleague, Luca Ventura, have published two papers and added *our* names as authors. That is the kind of academic camaraderie that makes everything you do worthwhile. And it makes up for the times I have seen papers published containing my work, but without any mention of my name. Hey, it happens. What're you gonna do?

Our enduring image of Gino is probably from the dinner we had with the city's mayor. If you watched a lot of *Mummy Road Show* episodes, you probably noticed that incredible dinners were one of the fringe benefits. And if you look at me, you can probably figure out that I enjoyed every

Ron and Gino work in close quarters to examine a well-dressed mummy in "Tales from an Italian Crypt."

last bite. Well, we were having a fantastic dinner, we had downed countless bottles of wine, and it was time for a toast. Gino whipped out a lung sample from one of the mummies and proposed that we toast it. And we did. Pure Gino.

There are so many people Ron and I want to remember and thank and recognize who worked with us on *Mummy Road Show* that it would be impossible to cover them all. One last person I would like to talk about, someone who deserves a book of her own, is Gretchen Warden, whom I have mentioned in previous chapters.

The bond I developed with Gretchen was kind of amazing. After I left Philadelphia for Panama City, Florida, in 1988, I would get calls from sideshow people that she had referred to me. I might get a message saying, "Grady Styles, the Lobster Boy, will be stopping by for X-rays." It was through Gretchen that I met Andrew Nelson, the physical anthropologist at UCLA, who was studying giants. I had x-rayed the skeleton of a giant at the Mütter Museum, and Andrew was very pleased with my work. When Andrew finished his dissertation and became a faculty member of the University of Western Ontario, he began working in Peru and asked me to come down to set up a field

X-ray site. My work in Peru led to the *Road Show* and other great stuff, all of which I can trace back to Gretchen.

I was really sad when she passed on. But I think probably because I spend so much time working on dead people, I have developed a sense that life is kind of this precarious thing anyway. We're only here for a really short time, so I think it is important to appreciate the time that you do have to spend with people. That hits home when I go up to the medical examiner's office to examine someone who has just died. Most of those people I see had no idea what was going to happen to them.

Gretchen was an exceptional person. I never met anyone else like her. Sometimes Ron and I got bogged down in the administrative stuff we have to do here, and we did not get to talk to her, but at least every couple of months we would get a phone call—or a bizarre message—from Gretchen. Once she left a message for me with my secretary that there was a fetal monster sale in North Carolina, and that she needed a ride. (Gretchen had never gotten a driver's license.) Another time she referred a performer, Harley Newman. He did a blockhead act, but instead of placing a spike up his nose, Harley used a Black & Decker drill, followed by spiked heels. Gretchen was always connecting with people like that.

Two for the Road Show

Ron and I have both been asked what we hope the legacy of *Mummy Road Show* will be. Obviously, the more people we can get excited about science, the better the chances are that someone will pick up the ball and run with it in the future. The great thing is that the program may already have had a very positive impact on the lives of at least two people we know of.

The first is Brian Finucane, who we met while filming "Mummy in a Closet." He had taken an interest in a Peruvian mummy of unclear origins at Cornell University, and we helped him work it up. He was a very intelligent young man and had very sound scientific methodology. We feel his work with us, and our connection to National Geographic Channel, helped him further distinguish himself, and he ended up earning a Rhodes Scholarship.

The second person was a student of ours, Kristen Horner. She was a brilliant young woman who was not used to working with groups of people. It was kind of her Achilles' heel at the time. She accompanied us on a couple of episodes, did some good research, got a feel for the group dynamic you have out in the field, and ultimately, this helped her get into a graduate program at Arizona State.

One of the goals of the series was to get people interested in science, and show them that things do not always work out the way you plan. Good scientists learn to adapt and retool, and are prepared to deal with a conclusion that may be totally different than the one they start with. I think one of the reasons Kristen got into grad school was that we provided her with access to research opportunities. She observed us and took off with her own project. I'm very proud of Kristen.

Ron and I did *Mummy Road Show* for three seasons. We had some unforgettable experiences and met some unforgettable people. We immersed ourselves in new cultures and old cultures, and expanded our palate to include everything from dog to llama to a lot of stuff we were never able to identify.

If you could ferment it, we probably drank it. If you could wrap it in a tortilla, we probably ate it. We picked up some good science along the way, and worked with some great scientists.

We felt lucky and honored to be able to tell the stories of so many people whose stories had previously gone untold. We separated fact from fiction when we could, and where the information was unclear or inconclusive, perhaps we laid the groundwork for future study.

Ironically, *Mummy Road Show* may have taught us more about the living than the dead. Going from country to country and culture to culture, we saw how an incredibly wide variety of people—from the poorest to the most educated and enlightened—interacted with mummies. I don't think anyone else in our field can make that claim.

Ron and I are so grateful for having had the opportunity to do *Mummy Road Show*. We accumulated so much valuable information, and made so many new friends. And with our memories being what they are, we're thankful all of it was caught on tape. Yeah, we joked around, but we were deadly serious about the science and the responsibility that came with it.

One thing that will always stay with us is the community of friends, colleagues, and fans that was created through working on the show. Hopefully, we have conveyed this feeling with the collaborative effort that brought about this book.

On a more personal level of accomplishment, *Mummy Road Show* became a showcase to present the

Ron and I were like kids in a candy store on *Mummy Road Show*.

skills of radiographers. For a long time, the general public has viewed the profession simply as one where we are the people who take X-rays. They believe that X-ray technicians just push buttons. Nothing could be further from the truth. Over the past thirty years, as imaging technology has become more sophisticated, the need for more skilled and educated operators has also increased. The profession has met the challenge. In fact, to signify developments in the field, X-ray technicians are now known as radiographers or radiological technologists. Hopefully, the series documented that radiographers have an ability to manipulate all the factors responsible for producing diagnostic images under less than perfect circumstances. For members of the profession working to produce the images that help the radiologists and other physicians make the diagnosis, the series showed that there are other non-medical applications for their skills. To my knowledge, it was the first series based partially on the abilities of a radiographer.

Let there be light! We loved improvising on *Mummy Road Show*.

Where to next? That's the fun of it—we just don't know.

As I've said before, Ron and I like the thought that we have no idea what we'll be doing a year from now. One thing is certain: We'll be searching for knowledge. Whether we're together or apart, studying mummies in Peru or Piscataway, or learning about something totally new, you can be sure we'll be into it knee-deep or elbow-deep, and loving every minute of it.

Appendix

JERRY ON RON

Since we finished work on *Mummy Road Show*, a lot of people have asked about the relationship Ron and I have. They assume that the closeness of our professional relationship is mirrored in our personal lives, that we finish work and then go hang out together. They always seem surprised when I tell them we rarely interact outside of work.

This actually says a lot about the professional relationship we have. Ron and I obviously get along really well and work together beautifully. We both have clinical backgrounds, so long before we met, each of us had embraced the idea of functioning as a member of a team. Still, I think back to some of the early *Mummy Road Show* episodes, and it's weird how he would start a sentence and I would finish it, or vice versa. We often seemed to be thinking the same thing at the same time on those shows, too, so it appeared we had been working as a team for years and years. The truth was, outside of doing a few presentations for kids, we were about as unpracticed and unpolished as two TV show hosts could get.

I take that back. Ron had some stage experience in his younger days, and he still is a tremendous musical talent. Actually, music has been an important bond between us. We both enjoy music, and there have definitely been times when it kind of helped us get through some difficult personal or professional stuff.

But never difficult stuff between the two of us. During our three-year run, I cannot recall ever having a really tense moment with Ron—or anyone else on the show, for that matter. Any tension viewers sensed on screen probably had to do with the time constraints that were placed on us. Sometimes the shooting schedule just got too tight, and I was not able to do everything I needed to. The producers would say, "Okay, Jerry, time to move on."

That could get kind of frustrating. To his credit, Ron always knew to give me a little extra room when this was happening. A great example was during "Cave Mummies of the Philippines," when we discovered that one of the mummies showed evidence of a caesarean section. That was mind-boggling to me. I could have spent three weeks on that mummy alone, but I was constantly reminded that this was not the focus of the show and that we had to keep the momentum going. Man, that was tough. Ron gave me a really wide berth, and I eventually chilled out.

Ron also was a great help when he saw that something was driving me nuts. We did not work for

Edward Meyer watches Ron work during the filming of "Carnival Mummy."

weeks and weeks on *Mummy Road Show* episodes—some of them were shot in just thirty-six hours. If something was bugging me going into the shoot (usually it was something related to my full-time job), there was always a chance I would carry it over into the episode. Ron had this wonderful way of keeping me centered so we could get to work. What's great is there were times when I would do the same for him. We were not professional TV people, so we couldn't just flick a switch and turn off what was happening in our lives. *Mummy Road Show* was true reality TV in that sense.

In general, I would say that Ron has a longer fuse than I do in annoying situations. When he gets agitated, though, it is fun to watch, especially when he gets enthusiastic about something. At these times he reminds me of Odie from *Garfield*. He almost trembles with excitement. When Ron gets this way, I always let him know it's okay, I have the tranquilizer darts.

I have so much respect for Ron. He is a good man, and his skills as a scientist are absolutely world-class. How can you not have a great relationship with someone like that?

RON ON JERRY

Jerry is a problem solver. He feeds on challenges. When we have worked together on a case and it is relatively straightforward, the big lip comes out and he practically starts pouting. If, on the other hand, we have to squeeze into a cave or drink cow's blood to get an image, Jerry comes alive. Jerry is also a good planner, and smart about working ahead.

The other thing that is incredible about Jerry is his work ethic. He is literally tireless. So many times during the filming of *Mummy Road Show*, Larry Engel would shout, "Break for lunch!" and Jerry would go, "No, no, no." I know where he was coming from—we had a finite amount of time to work and they practically had to pull us off a mummy sometimes. Both of us wanted to give everything we had.

One of the reasons Jerry and I work so well together is that, when I get stuck on something, he always seems to come up with a way to help me. By that I mean he could look at things from my perspective, specifically when it came to obstacles involving the endoscope on *Mummy Road Show*. And I like to think that sometimes when Jerry had a situation with his X-ray unit, I was able to do the same. We also have this ability to bounce things off each other and come up with solutions that neither of us might get on our own. We have a good sense of when to step in and contribute, and just as important, we know when to back off, give the other guy some space, or just leave him alone.

I know that there were times when we were both stressed out; if I sensed that he needed my input, I offered it. If I sensed that I should be working away from the volcano, then that is what I would do. We were both good

Jerry x-rays a fetus in "Muchas Mummies."

at stepping in and taking the lead when the other guy was distracted or upset about something, and we would do so without a word being exchanged. It was so natural that it was kind of scary.

Looking back, I cannot think of a time when it was anything but pleasant to work with Jerry. He is one of those classic thinkers, and the more I have gotten to know him, the more respect I have for the guy. He is a very rare individual.

LIVE ONES: PEOPLE WHO SHOULD BE MUMMIFIED IMMEDIATELY

JERRY
1. George Bush
2. Dick Cheney
3. Jerry Falwell
4. Rush Limbaugh
5. Jessica Simpson

RON
1. All Politicians
2. Bad Doctors
3. Bad Lawyers
4. Bad Teachers
5. Tie: Bad Dogs and Bad Musicians

WRAP STARS:
MUMMIES WE'D LIKE A CRACK AT

JERRY

1. Lenin
2. Evita Peron
3. Ho Chi Minh
4. Gold Tooth Jimmy
5. John St. Helen (reputed to be John Wilkes Booth)

RON

1. Galileo
2. Abraham Lincoln
3. Any saint
4. An Ibaloi mummy from the Kabayan jungle
5. Gold Tooth Jimmy

DYING TO KNOW MORE:
MUMMIES WE STILL HAVE SO MANY QUESTIONS ABOUT

JERRY

1. Princess Anna
2. Luang Pho Dang
3. Hazel Farris
4. Marie O'Day
5. Andy the Blockhead

RON

1. Sylvester
2. Hazel Farris
3. Luang Pho Dang
4. Princess Anna
5. The nobleman from Popoli

CHECK, PLEASE:
WEIRDEST MEALS WE'VE EATEN

JERRY

1. Lusta (fermented seal flipper)
2. Lacuda (roasted seal pup)
3. Cuy (pan-fried guinea pig)
4. Tripe, or grilled intestines (still not sure of the animal, but it was in Peru)
5. Seal testicles

RON

1. Face of a duck
2. Dirt soup
3. Barbecued llama
4. Bad goat
5. Dog

BURIED TREASURES:
MOVIES YOU COULD TAKE YOUR MUMMY TO

JERRY AND RON

1. *Bubba Ho-Tep* (Elvis and JFK battle a mummy at a nursing home—2004)
    "How can you go wrong?"

2. *The Mummy* (Boris Karloff—1932)
    "This is how our generation first learned about mummies."

3. *The Mummy* (Christopher Lee and Peter Cushing—1959)
    "What a great pair!"

4. *The Mummy* (Brendan Fraser—1999)
    "All the bells and whistles!"

5. *The Mummy's Ghost* (Lon Chaney Jr. and John Carradine—1944)
    "Carradine is a scary dude, even without special effects."

6. *The Mummy Returns* (Brendan Fraser—2001)
    "A solid sequel."

7. *Blood from the Mummy's Tomb* (Valerie Leon—1972)
    "The star is a Bond girl!"

8. *The Robot vs. The Aztec Mummy* (1959)
    "Part of Mexico's somewhat bizarre mummy movie tradition."

9. *Abbott and Costello Meet the Mummy* (1955)
    "A comedy classic."

10. *Lust in the Mummy's Tomb* (2000)
    "We couldn't come up with a tenth one, so we went to the video store. We found this in the room near the tanning booths. Don't take your Mummy to this one."

Note from Ron: When they make *Mummy Dearest: The Movie,* I think Sean Connery should play Jerry.

Note from Jerry: Wow, did he really say that? Then I think Jon Bon Jovi should play Ron. Once at a party, my teenage daughter got Bon Jovi to agree to pose with her for a photo, and just as I snapped the picture, Ron jumped in and threw his arm around him.

Credits:
The *Mummy Road Show* team
at Engel Brothers Media

EXECUTIVE PRODUCERS
Steve Engel
Larry Engel

SERIES PRODUCER
Mary Olive Smith

PRODUCERS, DIRECTORS AND WRITERS
Larry Engel
Mary Olive Smith
Amy Bucher
Whitney Wood
Alana Campbell
Heidi Burke
Meredith Fisher
Steve Engel

EDITORS
Ted Bourne
Ed Greene
Joanne Belonsky
Andey Ford
Charlotte Stobbs
Lora Hays
Matt Tassone

ASSOCIATE PRODUCERS AND SOUND RECORDISTS
Brigitte Bruylant
Heidi Burke
Alana Campbell
Kevin Carrigan
Meredith Fisher
Steve Flynn
Rachael Leiserson
Aysin Karaduman

CINEMATOGRAPHER
Larry Engel

PRODUCTION ASSISTANTS
Julie Almendral
Sirin Aysan
Jessica Beck
Steve Flynn
Melissa Hetzner
Joanie Hilliard
Aysin Karaduman
Christine Kiernan
Susie Landers
Rachael Leiserson
Michael Perusse
Jorge Peschiera

PRODUCTION MANAGER
Susan K. Lee

PRODUCTION AND POST COORDINATORS
Aysin Karaduman
Rachael Leiserson
Christine Kiernan
Joanie Hilliard
Julie Almendral
Sirin Aysan

GRAPHICS
Michael Perusse
Jorge Peschiera